Process Analytics

Seyed-Mehdi-Reza Beheshti • Boualem Benatallah •
Sherif Sakr • Daniela Grigori •
Hamid Reza Motahari-Nezhad • Moshe Chai Barukh •
Ahmed Gater • Seung Hwan Ryu

Process Analytics

Concepts and Techniques for Querying and Analyzing Process Data

 Springer

Seyed-Mehdi-Reza Beheshti
The University of New South Wales
Sydney, NSW, Australia

Sherif Sakr
The University of New South Wales
Sydney, NSW, Australia

King Saud bin Abdulaziz University
    for Health Sciences
Riyadh, Saudi Arabia

Hamid Reza Motahari-Nezhad
IBM Almaden Research Center
San José, CA, USA

Ahmed Gater
Ikayros
Paris, France

Boualem Benatallah
The University of New South Wales
Sydney, NSW, Australia

Daniela Grigori
Laboratoire LAMSADE
Université Paris Dauphine
Paris, France

Moshe Chai Barukh
The University of New South Wales
Sydney, NSW, Australia

Seung Hwan Ryu
The University of New South Wales
Sydney, NSW, Australia

ISBN 978-3-319-25036-6   ISBN 978-3-319-25037-3   (eBook)
DOI 10.1007/978-3-319-25037-3

Library of Congress Control Number: 2015956171

Printed on acid-free paper

This Springer imprint is published by Springer Nature
The registered company is Springer International Publishing AG Switzerland

Foreword

The quality of the services any organization provides largely depends on the quality of their processes. Companies are increasingly grasping this concept and moving towards process-oriented enterprises. As this happens, the central problem (and opportunity) for any organization becomes the identification, measurement, analysis, and improvement of its processes. Without this, it is very difficult—if not impossible—to be competitive.

This book provides researchers and practitioners with invaluable knowledge in the area of process management in general and process analysis in particular. The book starts from an introduction on process modeling and process paradigms, which is very easy to understand for novices but that also includes interesting bits of information for people who are experts in the area. From there, it proceeds to teaching how you can query and analyze process models and moves on to the analysis of process execution data. In this way, you get a complete picture of what you need to do to identify, understand, and improve business processes.

The book is very well written and a pleasure to read. It presents concepts that are normally not easy to grasp in a manner that is simple and intuitive and also a way to classify and structure topics that makes it easy to make sense out of a rather large set of concepts, paradigms, and techniques. All authors have many years of experience in Business Process Management (BPM) and, thanks to this and to a combined academia-industry background, they have been clearly able to identify which are the key, hard analytics problems in BPM that are relevant in practice and they guide the readers to show them how these problems can be solved concretely.

What I particularly enjoyed about reading this book, and I am sure you will enjoy as well, is that it deals with very realistic environments, where analysis is complex: in any company, you have very different systems and very different data, all related to the same processes. Indeed, most organizations have accumulated systems that support different subprocesses over the years. This means that if you really want to understand what is going on in your business and how to make it better, you need to get to the individual data source, correlate the data, follow the breadcrumbs, and then work up to discover the process model and its key performance indicators. Performing all of this, potentially in an explorative manner, needs process data querying, data correlation, data analysis, as well as process model-level analysis and matching techniques described in this book. Through such a journey, you will have a clearer picture of how much each process takes in terms of time, resources, and cost and where the "bottlenecks" are with regard to efficiency and effectiveness, which will help understand where and how to improve. This is an aspect that many authors forget or neglect, but with this book you will learn relatively simple but effective ways of approaching such scenarios.

I hope you like the book as much as I did. Enjoy your reading!

Trento, Italy   Fabio Casati
2015

Preface

Business processes are inseparable from data—data from the execution of business processes, documentation and description of processes, process models, process variants, artifacts related to business processes, and data generated or exchanged during process execution. Process data can be in various forms from structured to semistructured and unstructured. A variety of data capturing, collection, and instrumentation tools and process implementation over various types of systems have amplified the amount of data that are gathered and are available about processes. Business process monitoring, analysis, and intelligence techniques have been investigated in depth over the last decade to enable delivering process insight and support analytic-driven decision making about processes. However, they have been developed under the assumption of the existence of a workflow engine as the central repository for process data and for process analysis. As processes touch almost every system and leave an operational data footprint everywhere in an enterprise, and the wide span of process support the majority of information and enterprise application systems, the existing tools and methods have proved to be inadequate and limiting for correlating, managing, and analyzing process data.

Considering various process analytics needs and the importance of process analytics over a myriad of process data forms, formats, and levels of abstractions, we felt the need for writing this book as a technical introduction to the field of process analytics to share the state of the art in research achievements and practical techniques of process analytics. Given that many "data analysis" subjects such as querying, matching, and warehousing are covered well in the literature, the focus of this book will be on concepts, techniques, and methods for analyzing process data and offering a common understanding of the concepts and technologies in process analytics. This book will cover a large body of knowledge in process analytics, including process data querying, analysis, and matching and correlating process data and models to assist practitioners as well as researchers in understanding underlying concepts, problems, methods, tools, and techniques for modern process analytics. It starts by introducing basic business process and process analytics concepts, describes the state of the art in this space, and continues by taking a deeper perspective on various analytics techniques covering analytics over different levels of process abstractions, from process execution data to methods for linking and correlating process execution data and to inferring process models and querying process execution data and process models and scalable process data analytics methods. The book also provides a review of commercial process analytics tools and their practical applications.

Chapters Overview

In particular, we introduce concise and commonly accepted definitions, taxonomies, and frameworks for business process management in Chap. 2. We briefly discuss process-centered systems, process modeling, and why process model developers have to model collaboratively in today's world. We classify strategies that a business may use to automate processes. We discuss technologies, applications, and practices used to provide business analytics. We discuss a wide spectrum of business process paradigms that has been presented in the literature from structured to unstructured processes. We discuss state-of-the-art technologies and demystify the concepts, abstractions, and methods in structured and unstructured BPM, including activity-, rule-, artifact-, and case-based processes.

Chapter 3 presents process matching techniques that are useful in a series of process model analytics tasks. We present a brief overview of schema matching techniques as they are useful for organizing process data, more precisely for the integration of process execution data from various, potentially heterogeneous, systems and services in a process event log. Besides integrating process execution data, process models form various repositories (or resulting from process mining tasks) have to be assembled in the process model layer of the process space (see Fig. 1.3). The management of such integrated repositories requires techniques to manage versions (operators like merge, diff) and for duplicate detections that are based on process matching. We also classify approaches for matching different dimensions of process models (data, interface, protocol, etc.) and analyze their strengths and limitations. We also discuss that process matching techniques are needed for conformance evaluation (comparing the process model mined from event logs and the initially designed process model or its description in textual documentation). Process similarity measures presented in Chap. 3 are also used for clustering process models and for organizing process models mined from event logs for unstructured processes, like those in the medical domain.

Querying is an important aspect in business process analytics. In particular, querying techniques are commonly used in the different phases of business process life cycles for various analytics purposes. For example, during the design time, querying techniques can be used to retrieve existing process models, check the similarity between new process models and existing processes, and conduct compliance checks against existing business logic or rules. During the runtime phase of the business process life cycles, querying techniques can be utilized for analyzing and understanding the execution patterns of business processes. In addition, they can be used to instantly detect any compliance violations. In offline mode, querying techniques can be used for auditing, mining, and process improvement purposes. To address the importance of querying business processes, in Chap. 4 we focus on business process querying techniques and languages. We describe the key concepts, methods, and languages for querying business processes. We describe the foundation and enabling technology for understanding the execution of a business process in terms of its scope and details, which is a challenging task and requires querying process repositories and business process execution.

Identifying business needs and determining solutions to business problems require the analysis of business process data. In Chap. 5, we discuss that the analysis of business data will help in discovering useful information, suggesting conclusions, and supporting decision making for enterprises. In Chap. 5, we give an overview of the different aspects of business data analysis techniques and approaches from process/data spaces to data provenance and data-based querying techniques. We provide an overview of *warehousing process data* followed by an introduction to *data services* and *DataSpaces*, which facilitate organizing and analyzing process-related data. We discuss the importance of supporting big data analytics over process execution data. We then take a holistic view of the process executions over various information systems and services (i.e., process space) followed by a brief overview of process mining to highlight the interpretation of the information in the enterprise in the context of process mining. Moreover, we focus on introducing cross-cutting aspects in process data and discuss how process analytics can benefit from cross-cutting aspects such as provenance, e.g., to analyze the evolution of business artifacts.

Finally, Chap. 6 provides an overview of open-source and commercial software for process analytics. Software for process analytics can be applied to the rich source of events that document the execution of processes and activities within BPM systems in order to support decision-making in organizations. In this context, various existing tools focus on the behavior of completed processes, evaluate currently running process instances, or focus on predicting the behavior of process instances in the future. In this chapter, we provide a summary and comparison of existing open-source and commercial software for process analytics, including real-world use case scenarios, followed by a discussion and future directions on some of the emerging and hot trends in the business process management area such as process spaces, big data for processes, crowdsourcing, social BPM, and process management on the cloud. We briefly describe the core essence of these directions and discuss how they can facilitate the analysis of business processes.

Who Is This Book For?

In writing this book, we have taken into consideration a wide range of audience interested in the space of business process management and process analytics. We have tried to cover topics of interest for academics (professors, researchers, and research students), professionals (managers, data scientists, analysts, and software engineers), and practitioners in understanding and employing process analytics methods and tools to gain a better insight within and across business processes. This book is a comprehensive textbook on process analytics and, therefore, could be a useful reference for academics, professionals, and practitioners.

To Professors You will find this book useful for a variety of courses, from an undergraduate course in business process management up through a graduate course in business process analytics. We have provided considerably more material than can fit in a typical one-term course; therefore, you can think of the book as a buffet from which you can pick and choose the material that best supports the course you wish to teach.

To Research Students and Researchers We hope that this textbook provides you with an enjoyable introduction to the field of process analytics and business process management. We have attempted to properly classify the state of the art, describe technical problems and techniques/methods in depth, and highlight future research directions in process analytics.

To Professionals and Practitioners You will find this book useful as it provides a review of the state of the art in commercial tools and techniques and also describes real-world use case scenarios. The wide range of topics covered in this book makes it an excellent handbook on process analytics. Most of the methods that we discuss in each chapter have great practical utility: process matching techniques in Chap. 3, process querying techniques and languages in Chap. 4, and process data analysis in Chap. 5. Therefore, in each chapter you can find details for the state-of-the-art tools and methods. An overview of open-source and commercial software for process analytics is finally provided in Chap. 6.

Sydney, New South Wales, Australia	Seyed-Mehdi-Reza Beheshti
Sydney, New South Wales, Australia	Boualem Benatallah
Sydney, New South Wales, Australia	Sherif Sakr
Paris, France	Daniela Grigori
San José, California, USA	Hamid Reza Motahari-Nezhad
Sydney, New South Wales, Australia	Moshe Chai Barukh
Paris, France	Ahmed Gater
Sydney, New South Wales, Australia	Seung Hwan Ryu

Contents

List of Figures

Chapter 1
Introduction

Business processes are central to the operation of public and private enterprises. For most enterprises, the success is strictly related to how efficient and effective the execution of their processes is. For this reason, business process analytics has always been a key endeavor for companies. Early efforts to address this goal started with process *automation* where workflow and other middleware technologies were used to reduce human involvement by better systems integration and automated execution of business logic. The total or partial automation of the process creates an unprecedented opportunity to gain visibility on process executions in the enterprise. Recently, the focus of process thinking has shifted toward understanding and analyzing business processes and business process-related data captured in various information systems and services that support processes. In this chapter, we give an overview of the different approaches for analyzing business processes at the different phases of its life cycle.

1.1 The Modern Enterprises and the Need for Process Analytics

Information processing using knowledge-, service-, and cloud-based systems has become the foundation of twenty-first-century life. These systems run and support processes in our governments, industries, transportations, and hospitals and even our social life. In this context, business processes (BPs) and their continuous improvements are key to the operation of any system supporting our life and in enterprises. Typical examples of BPs that are supported by systems include those that automate the operation of commercial enterprises such as banking and financial transaction processing systems.

Over the last decade, many BPs across and beyond the enterprise boundaries have been integrated. Process data is stored across different systems, applications, and services in the enterprise and sometimes shared between different enterprises

© Springer International Publishing Switzerland 2016
S.-M.-R. Beheshti et al., *Process Analytics*, DOI 10.1007/978-3-319-25037-3_1

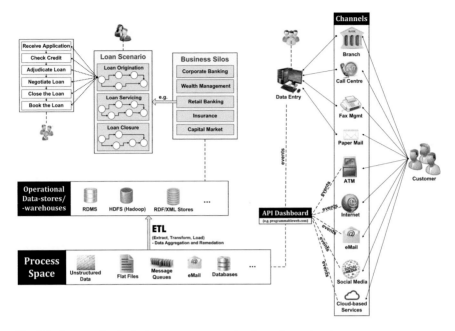

Fig. 1.1 An example of a business process execution in modern enterprises

to provide the foundation for business collaborations. Figure 1.1 illustrates a
business process scenario in the banking context that spans across multiple systems
and services inside and among third-party services providers. These systems are
distributed over various networks, but when viewed at a macrolevel, various
organizations and systems are components of a larger, logically coherent system.
Let us have a closer look at this example scenario.

Figure 1.1 demonstrates a property lending scenario in the banking system.
Consider Adam (a customer) who plans to buy a property. He needs to find a lending
bank. He can use various crowdsourcing services (e.g., Amazon Mechanical Turk[1])
or visit a mortgage bank to find candidate banks. Then he needs to contact the banks
through one of many channels and start the loan pre-approval process. After that, he
may visit various Web sites or real-estate services to find a property. Meanwhile, he
can use social network Web sites (e.g., Facebook or Twitter) to socialize the problem
of buying a property and ask for others' opinions on finding a suburb. After finding
a property, he needs to get a solicitor to start the process of buying the property.
Lots of other processes can be executed before or after this point. For example, the
bank may outsource the process of confirming Adam's income, during pre-approval
process, to other companies or appraising the property.

---

[1] https://www.mturk.com/

As can be seen, the data relevant to the business process of a bank is scattered across multiple systems, and in many situations, stakeholders can be aware of processes, but they are not able to track or understand it: it is important to maintain the vital signs of bank processes by analyzing the process data. This task is challenging as a huge amount of data needs to be processed (it requires scalable methods) and enables answering process-centric queries, e.g., "Where is loan-order #756? What happened to it? Is it blocked? Give me all the documents and information related to the processing of loan-order #756. What is the typical path of loan orders? What is the process flow for it? What are the dependencies between loan applications A_1 and A_2? How much time and resources are spent in processing loan orders that are eventually rejected? Can I replace X (i.e., a service or a person) by Y? Where did the data come from? How was it generated? Who was involved in processing file X? At which stage do loan orders get rejected? How many loan orders are rejected between the time period τ_1 and τ_2? What is the average time spent on processing loan orders? Where do delays occur?" Can the bank outsource the process of confirming Adam's income or appraising the property to provider X (i.e., is the business protocol of the provider X compatible with the process model of the bank)? What are the differences between public process models of two different banks?

Under such conditions, *process analytics* becomes of a great practical value but clearly a very challenging task as well.

1.2   Business Processes: An Overview

A business process (BP) is a set of coordinated tasks and activities, carried out manually or automatically, to achieve a business objective or goal [6, 346]. An activity is the smallest unit of work performed by executing a program, enacting a human or machine action, or invoking another business process (known as subprocess) [7, 346]. A business process is typically *structured* and, therefore, is associated with a data flow, showing what data and how they are transferred among activities and tasks, and a control flow, showing the order in which activities and tasks are performed.

Two types of business processes are recognized: public and private. Public business processes can be shared with business partners (e.g., clients and suppliers) within an enterprise and can be used in the business-to-business integration (B2Bi) context [29]. On the contrary, private business processes are internal to the enterprise, include execution details, and can be used in enterprise application integration (EAI) context [29]. In order to manage organization performance through BPs, a set of methods, techniques, and tools are defined, developed, and presented under a concept known as business process management (BPM).

A BPM system is a "generic software system that is driven by explicit process designs to enact and manage operational business processes" [7], where operational processes refer to repetitive business processes performed by organizations in

their daily operations, which are explicitly defined and modeled, e.g., the product
shipping process. In this context, processes at the strategic decision-making level,
which are performed by the enterprise management, or processes that are not explicit
are excluded [6, 346]. For instance, if part of the business process is performed by
some legacy systems or services that their processes are not explicitly defined, they
are not covered.

In a more generic and broader context, the notion of a process-aware information system (PAIS) [1] refers to a software system that manages and supports
the execution of operational processes involving people, applications, and other
information sources, containing process-related data, and is centered around explicit
representation of process models. Still, there are a large body of information systems
within the enterprise which support (at least part of) a process; however, they do
not have an explicit representation of process models. The data in such systems
may contain information related to several versions (evolution) of the process,
as process is adapted and updated during system operation due to frequent and
unprecedented changes in the business environment. Enabling the understanding
of business process in such a hybrid environment needs not only ability to analyze
processes in systems with explicit process representation but also in those without
such an explicit representation. This means enabling end users to issue queries to
monitor, analyze, and understand the execution information of business processes
and related business artifacts.

Today, BPM systems (BPMSs) enable organizations to be more efficient and
capable of process automation throughout the process management *life cycle*. The
BPM life cycle can be divided into four phases [6]: (a) *design*—in this phase, the
process is (re-)designed and modeled, where the designs are often graphical and the
focus is on "structured processes" that need to handle many cases [7]; (b) *configuration*—during this phase, a process-aware system, e.g., a workflow management
system (WfMS), is configured; (c) *enactment*—in the process enactment phase, the
operational business process is executed; and (d) *diagnosis*—in the diagnoses phase,
the process is monitored and analyzed, and process improvement approaches are
proposed.

In the design phase, a modeling tool or technique [236] (e.g., BPMN, CogNIAM,
xBML, and EPC) will be used to specify the order of tasks in the business process.
Process modeling tools typically support a graph-based modeling approach adopting
a popular process modeling notation such as the business process modeling notation
(BPMN) [365]. Process models created in the process design phase are usually
too high level to be executed, and therefore, an analytical process model must
be configured to an executable process model. Moreover, BPM suite software
provides programming interfaces (e.g., BPEL [161], WS-CDL,[2] and XPDL or other
technologies including model-driven/service-oriented architecture) which allow
enterprise applications to be built to leverage the BPM engine. Workflows and
WfMSs were introduced to support BPs' life cycle.

---

[2]http://www.w3.org/TR/ws-cdl-10

The Workflow Management Coalition [116, 227] defines a workflow as "the automation of a business process, in whole or part, during which documents, information or tasks are passed from one participant to another for action, according to a set of procedural rules" and defines a WfMS as "a system that defines, creates and manages the execution of workflows through the use of software, running on one or more workflow engines, which is able to interpret the process definition, interact with workflow participants, and where required, invoke the use of IT tools and applications."

From the definition it can be seen that a WfMS only supports the BPM life cycle from the process design to the process enactment phases. To expand the scope of WfMSs, BPMSs were introduced as an extension of classical WfMSs focusing more on the diagnosis phase of the BPM life cycle, i.e., monitoring, tracking, analysis, and predication of business processes in systems with explicit process models. In a BPMS, various resource classes, e.g., human or nonhuman, application or non-application, and individual or teamwork, could contribute to tasks within a business process. Most of BPMSs focus on task-resource allocation in terms of individual human resources only [81, 369], i.e., they model control flow dependencies between activities in structured business processes.

Currently, many workflow vendors are positioning their systems as BPMSs. In particular, WfMSs [450] (e.g., Staffware, MQSeries, and COSA) can be used to integrate existing applications and support process changes by merely changing the workflow diagram [6, 395, 450]. Isolating the management of business processes in a separate component is also consistent with recent developments in the domain of Web services, where many information systems in the enterprise have been implemented using Web services. Web service technology has become the preferred implementation technology for realizing the service-oriented architecture (SOA) paradigm [29, 370]. SOA is an architectural style that provides guidelines on how services are described, discovered, and used. In particular, software applications are packaged as "services," where services are defined to be standard based and platform and protocol independent to address interactions in heterogeneous environments [29, 370].

Web service composition languages such as BPEL4WS, BPML, WSCI, XLANG, and WSFL can be used to glue services defined using WSDL[3] together. Web services provide standard specifications to simplify integration at lower levels of abstractions [284], i.e., messaging, or at higher abstraction levels [344–346], i.e., service interfaces, business protocols, and also policies. In fact, what is provided at the higher levels of abstractions are languages to define the service specifications, i.e., its interface, business protocol, and policies. Enabling the analysis of service interactions, in the context of business process executions, and that of service integration is a goal of enterprises today. In particular, business process analysis

---

[3]Proposed by IBM and Microsoft, and later published as a W3C note [404], WSDL is a general-purpose XML language for describing the interface and also the implementation of Web services. The interface describes the operations that a service offers.

(BPA) over a wide range of information systems, services, and software that implement the actual business processes of enterprises is required.

BPA is particularly concerned with the behavioral properties of enacted processes, e.g., at runtime, through *monitoring* BPs, or after execution, through using process *mining* or *querying* techniques [375]. BPA is typically structured around three different views [347]:

- *Process view* is concerned with the enactment of processes and is thus mainly focused on the compliance of executed processes. Process view is the core view of process monitoring and controlling. Key performance indicators (KPIs) related to the business processes can be evaluated using dimensions such as workflow and activity models and considering measures such as time, cost, or quality. Corresponding attributes in the process models can be updated continuously using the result gained within this view.
- *Resource view* is centered around the usage of resources within processes. The resource view is dealing with the resource performance to analyze the usage of resources within processes, to analyze the costs/time/quality of the resources, and to detect situations where high process efficiency is accompanied by a poor usage of the available resources. The quality of a resource can be measured using efficiency and effectiveness (total output).
- *Object view* focuses on business objects, e.g., inquiries, orders, or claims and analyzes the life cycle of these objects. The object view considers business objects as first-class citizens and analyze the life cycle of these objects using cost (e.g., costs for the handling of an object), time (e.g., typical processing time for these objects), and quality (e.g., potential problems related to an object).

These three views are populated with statistical information (e.g., the minimum, the average, or the deviation of some parameter of interest), and correlations are typically established across them. It is important as differentiating these three views can reduce the complexity for process monitoring and controlling. In this context, every view has its own purpose, but various interrelations exist between the three views. Designing and maintaining BPA applications is challenging, as business processes in modern enterprises are developed by different communities of practice, reside on different levels of abstractions, and are handled by different IT systems [174].

It should be noticed that, besides monitoring KPIs and quantitative data, it is possible to monitor the compliance of business processes with relevant regulations, constraints, and rules during runtime [129, 305, 312]. For example, it is possible to use a runtime verification framework based on linear temporal logic and colored automata [312] and use properties expressed in linear-time temporal logic (LTL) to deal with rules and expectations [129]. In this context, monitoring will include continuously observing possible compliance violations and also the ability to provide fine-grained feedback and to predict possible compliance violations in the future [305].

1.2.1 Business Processes in Modern Enterprises

The business world is getting increasingly dynamic as various technologies such as social media and Web 2.0 have made dynamic processes more prevalent. For example, email communication about a process, instant messaging to get a response to a process-related question, allowing business users to generate processes, and allowing frontline workers to update process knowledge (using new technologies such as process wikis) [432] make the use of complex, dynamic, and often knowledge-intensive activities an inevitable task.

Such ad hoc processes have flexible underlying process definition where the control flow between activities cannot be modeled in advance but simply occurs during runtime [150, 155, 425]. In such cases, the process execution path can change in a dynamic and ad hoc manner due to changing business requirements, dynamic customer needs, and people's growing skills. Examples of this are the processes in the area of government, law enforcement, financial services, and telecommunications. In this book, we use the term ad hoc to refer to this category of processes.

Existing BPM tools (WfMSs and BPMSs) support well-structured [395] processes and do not provide sufficient flexibility to reflect the nature of ad hoc processes: structured processes are fully prescribed how a future decision will be made. Ad hoc processes can be divided into two types: unstructured and semi-structured. An unstructured process is a process that cannot be reduced to well-defined rules [425], unlike well-structured processes. A semi-structured process, or *case-based* process, is a process which contains both structured and unstructured subprocesses [396].

While PAISs do a great job in increasing the productivity of organizations, it is known that their rigidity restricts their applicability [179, 400, 425]. A number of challenges for the next-generation BPM have been discussed in [400]. For example, generation, recognition, and application of reusable "task patterns" and "process patterns" are suggested as an alternative to static workflows. Best practice processes [343] can be considered as a related approach, where the focus is to recommend activity steps in collaborative IT support systems. It is important as capturing and sharing the knowledge of how previous similar cases have been resolved becomes useful in recommending what steps to take and what experts to consult to handle a new case effectively.

Basic directions for the utilization of task-based approaches, to support users engaged in intensive and unstructured knowledge work, have been discussed in [192]. The gap between completely ad hoc processes and rigid, predefined business processes have been discussed in [73]. An extended state-of-the-art study in the area of flexible workflows and task management and a further approach for integrating ad hoc and routine work are presented in [247]. These approaches provide frameworks for enabling delivery of process models, process fragments, and past cases for tasks where different stakeholders can enrich task resources and information. Moreover, they reveal major issues concerning business process flexibility and how it can be facilitated through interactive processes models [425].

There are various solutions for organizing and managing the ad hoc nature of unstructured and semi-structured processes, ranging from document-based, task-based, and case-based techniques [9, 229]. They proposed to use document-based, task-based, and case-based techniques to manage the ad hoc nature of such processes. Advanced techniques for building personal knowledge spaces and wiki-based collaborative document spaces are also discussed in these approaches. Some other techniques used email, which plays a central role for the exchange of tasks and task-related information in organizations [64]. Also it is possible to support agile business processes focusing on email-based and human-to-human cooperation, where the collaboration flow determines the enterprise process flow [425].

Ad hoc processes are complex not only because they are scattered across several systems and organizations, but also they require many different people (having lots of knowledge and experience) to collaborate to find the correct solution. *Case management*, also known as case handling [433], is defined as a common approach to support knowledge-intensive processes. Moreover, various technologies, e.g., customer relationship management (CRM) and content management systems (CMS), have been proposed for supporting such dynamic and ad hoc processes; however, they are not sufficient to address the key requirements of these types of processes: they are primarily driven by human participants reacting to changing context and do not follow a predetermined path.

To understand the problem, Fig. 1.2 illustrates an example scenario in the domain of semi-structured (case-based) processes. This scenario is based on breast cancer treatment cases in Velindre hospital [433]. In this scenario, a general practitioner (GP), suspecting a patient has cancer, updates the patient history using hospital information system and refers the patient to a breast cancer clinic (BCC). BCC checks the patient's history and requests assessments such as an examination,

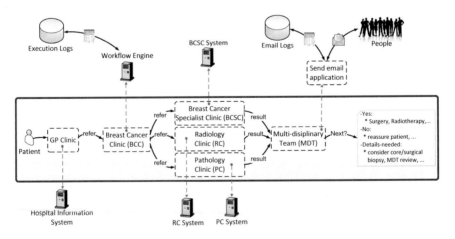

Fig. 1.2 An example of the ad-hoc business process execution in an enterprise

imaging, fine-needle aspiration, and core biopsy using a workflow engine. Therefore, the workflow system refers the patient to a breast cancer specialist clinic (BCSC), radiology clinic (RC), and pathology clinic (PC), where these departments use their own systems, apply medical examinations, and send the results to the multidisciplinary team (MDT). The results are gathered and sent to MDT members (e.g., surgeon oncologist, radiologist, pathologist, clinical and medical oncologist, and nurse) through a send mail application. Analyzing the results and the patient history, MDT members will decide for the next steps, e.g., in case of positive findings, nonsurgical (radiotherapy, chemotherapy, endocrine therapy, biological therapy, or bisphosphonates) and/or surgical options will be considered.

As illustrated in the scenario, most of the processes are conducted in an ad hoc manner and do not rely on integrated software solutions. Instead, a mix of computerized systems and direct collaborations are used. For instance, the processing of orders in the hospital information system may involve the accounting system and personal productivity tools of the laboratories to follow up patients. Hence, the data relevant to a business process is then scattered across multiple systems, e.g., RC, BCSC, and PC system, with no integration between them, or they are spread across multiple documents stored in the personal folders of GPs and exchanged by communication tools such as emails. Moreover, the process itself might be only partially specified or not specified at all. This yields to the fact that in many situations, stakeholders can be aware of processes, but they are not able to track or understand it. Therefore, under such conditions, *organizing*, *indexing*, and *querying* ad hoc process data become of a great practical value but clearly a very challenging task as well.

In such scenarios, organizations (e.g., health and medicine) create and run hundreds of business process models to support the ad hoc nature of such processes. In this context, matching between two workflow processes is the key step of (workflow) process reuse; however, in the motivated scenario, flexible workflows require inexact matching and querying approaches to support the ad hoc nature of such processes. To address this challenge, repositories have been proposed for managing large collection of process models, offering a rich set of functionalities, such as storing, querying, updating, and versioning process models. From the querying point of view, it will be important to query process models (support for finding process models) and the process-related data. In practice, business process models are not built just for communication purposes among business experts. Processes are executable. In addition to proprietary workflow languages, standardized executable processes exist. For example, the Business Process Execution Language for Web Services (BPEL4WS) is a standardized executable language [125] with major software vendors supporting it. There are many scenarios where the execution of these processes needs to be monitored and analyzed. Therefore, querying techniques for the business process executions serve as fundamental components to realize the monitoring and analytics goals.

Most related work in the area of analyzing business process execution assumes well-defined processes, and the current state of the art in querying business processes does not provide sufficient techniques for the analysis of ad hoc process data.

For example, some of the basic assumptions of existing BP querying techniques are that each event should be registered in the log, the BP models should be predefined and available, and the execution traces should comply with the defined process models. In case of ad hoc processes, the understanding of processes and analyzing BP execution data are difficult due to the lack of documentation and especially as the process scope and how process events across these systems are correlated into process instances are subjective: depending on the perspective of the process analyst.

In this context, most objects and data in the process execution data are interconnected, forming complex, heterogeneous but often semi-structured networks, and can be modeled using graphs. Understanding modern business processes entails identifying the relationships among entities in process graphs. Viewing process logs as a network and studying systematically the methods for mining such networks of events, actors, and process artifacts are a promising frontier in database and data mining research: process mining provides an important bridge between data mining and business process modeling and analysis [11]. In particular, process mining subsumes process analytics and enhances it with more comprehensive insights on the process execution: process mining techniques can be used to identify bottlenecks and critical points through "replaying" the execution traces used to discover a process model and *enrich* the discovered model with quantitative information.

There are many studies on the analysis of graphs, such as network measures [463], statistical behavior study [355], modeling of trend and dynamic and temporal evolution of networks [228, 268], clustering [387], ranking [430], and similarity search. All these approaches can be leveraged for mining and analyzing process graphs. Moreover, for effective discovery of ad hoc process knowledge, it is important to enhance process data by various data mining methods, i.e., to help data cleaning/integration, trustworthiness analysis, role discovery, and ontology discovery, which in turn help improving business processes.

To address these challenges, a set of works [346, 403] focused on the correlation discovery between events in process logs, i.e., event correlation is the process of finding relationships between events that belong to the same process execution instance. In particular, the problem of event correlation can be seen as related to that of discovering functional dependency [299, 380] in databases. These works are complementary to process mining techniques as they enable grouping events in the log into process instances that are then input to process mining algorithms.

Another solution is to convert process execution data into knowledge to support the decision-making process. To achieve this a family of methods and tools have been proposed for developing new insights and understanding of business performance based on collection, organization, analysis, interpretation, and presentation of ad hoc process data [38, 265, 331]. Another line of related work [56, 57, 60, 61], focused on providing a framework, simple abstractions, and a language for the explorative querying and understanding of process graphs from various user perspectives. The proposed framework caters for life cycle activities important for wide range of processes, from unstructured to structured, including understanding, analyzing, correlating, querying, and exploring process execution data in an interactive manner.

1.3   Process Analytics

The focus of process improvement activities in the enterprise has shifted to process analytics, where the goal is to understand how a business process is performed, at the first place, and then to identify opportunities for improvement. However, the wide-scale automation in enterprises has led to having business processes implemented over many systems. Therefore, answering questions such as "Where is order #*X*? What happened to it? At which stage do orders get rejected? How much time and resources are spent in processing orders? Where the file #*Y* came from? How it was generated? Who was involved in updating this file?" becomes difficult at best.

The main barrier for answering questions like those above is that in modern enterprises the information about process execution is scattered across several systems and data sources. Consequently, process logs increasingly come to show all typical properties of the *big data* [483]: wide physical distribution, diversity of formats, nonstandard data models, and independently managed and heterogeneous semantics. We use the term *process data* to refer to such large hybrid collections of heterogeneous and partially unstructured process-related execution data.

Digitalization of business artifacts such as documents and reports, generating huge metadata such as versioning and provenance [58] for imbuing business artifacts with additional semantics, and adoption of social media (e.g., social BPM [432]) will generate part of the process data. Apart from the business-related data, various types of metadata (such as versioning, provenance, security, and privacy) may be collected on several systems and organizations, especially as processes require many different people (having lots of knowledge and experience) to collaborate to find the correct solution. Process analytics begins with understanding the process data. In this context, before analyzing and querying your process data, there is a need to capture and organize the process data.

1.3.1   Capturing Process Data

Multiple methods are available for capturing process data. In the process domain, data services play an important role in capturing process-related data [98, 99, 202, 399]. For example, when an enterprise wishes to controllably share data (e.g., structured data such as relational tables, semi-structured information such as XML documents, and unstructured information such as commercial data from online business sources) with its business partners, via the Internet, it can use data services to provide mechanisms to find out which data can be accessed, what are the semantics of the data, and how the data can be integrated from multiple enterprises.

In particular, data services are "software components that address these issues by providing rich metadata, expressive languages, and APIs for service consumers to send queries and receive data from service providers" [99].

Data as a service, or DaaS, is based on the concept that the data can be provided on demand to the user regardless of geographic or organizational separation of provider and consumer [445]. In particular, data services are created to integrate as well as to service enable a collection of data sources. These services can be used in mash-ups, i.e., Web applications that are developed starting from contents and services available online, to use and combine data from two or more sources to create new services. In particular, data services will be integral for designing, building, and maintaining SOA applications [98]. For example, Oracle's ODSI supports the creation and publishing of collections of interrelated data services, similar to *dataspaces*.

Dataspaces are an abstraction in data management that aim to manage a large number of diverse interrelated data sources in enterprises in a convenient, integrated, and principled fashion. Dataspaces are different from data integration approaches in a way that they provide base functionality over all data sources, regardless of how integrated they are. For example, a dataspace can provide keyword search over its data sources, then more sophisticated operations (e.g., mining and monitoring certain sources) can be applied to queried sources in an incremental, pay-as-you-go fashion [206]. These approaches do not consider the business process aspects per se; however, they can be leveraged for organizing and managing ad hoc process data. DataSpace Support Platforms (DSSPs) have been introduced as a key agenda for the data management field and to provide data integration and querying capabilities on (semi-)structured data sources in an enterprise [206, 413]. For example, SEMEX [94] and Haystack [252] systems extract personal information from desktop data sources into a repository and represent that information as a graph structure where nodes denote personal data objects and edges denote relationships among them.

Recently, a new class of data services has been designed for providing data management in the cloud [459]: the cloud is quickly becoming a new universal platform for data storage and management. In practice, data warehousing, partitioning, and replication are well-known strategies to achieve the availability, scalability, and performance improvement goals in the distributed data management world. Moreover, database as a service is proposed as an emerging paradigm for data management in which a third-party service provider hosts a database as a service [202]. Data services can be employed on top of such cloud-based storage systems to address challenges such as availability, scalability, elasticity, load balancing, fault tolerance, and heterogenous environments in data services. For example, Amazon Simple Storage Service (S3) is an online public storage Web service offered by Amazon Web Services.[4]

---

[4]http://aws.amazon.com/

1.3.2 Organizing Process Data

Organizing process data requires arranging data in a coherent form and to systematize its retrieval and processing. In today's modern enterprises, process-related data come from many sources and can be used for multiple purposes. The process data inputs are very large and need to be organized and stored for computer processing. This task is challenging as (over time) new data sources are becoming accessible, data streams change, and data usage (how we use and process data) evolves. Even historical data is not necessarily static and unchanging. Another challenge is the size of data, where there is a need for big data platforms and techniques to process huge amounts of unstructured process data.

In order to organize the process-related data, the first step is gathering and integration of process execution data in a *process event log* from various, potentially heterogeneous, systems and services. In general, this step involves several phases [225, 391]: (a) data analysis is required to detect errors and inconsistencies of heterogeneous event logs; (b) definition of transformation workflow and mapping rules, depending on the number of data sources and their degree of heterogeneity of the data, a large number of data transformation and cleaning steps may have to be executed; (c) verification, the correctness and effectiveness of a transformation workflow/definitions should be tested and evaluated; and (d) transformation, data transformations deal with schema/data translation and integration and with filtering and aggregating data to be stored in the integrated process log.

The next step is providing techniques to identify entities (e.g., process stakeholders and process artifacts) and the interactions among them within such integrated process logs. In this context, most entities (structured or unstructured) in process logs are interconnected through rich semantic information, where entities and relationships among them can be modeled using graphs. Since graphs form a complex and expressive data type, there is a need for methods to organize and index the graph data. Existing database models, including the relational model, lack native support for advanced data structures such as graphs. In particular, as the graph data increases in size and complexity, it becomes important that it is managed by a database system. There are several approaches for managing graphs in a database. A line of related work extended a commercial RDBMS engine, e.g., Oracle provides a commercial DBMS for modeling and manipulating graph data [18], and some other works [59] used general-purpose relational tables to support graph structure data, e.g., triplestore which is a special-purpose database for the storage and retrieval of RDF [315] (Resource Description Framework).

A new stream of work used MapReduce [131] for processing huge amounts of unstructured data in a massively parallel way. Hadoop [470], the open-source implementation of MapReduce, provides a distributed file system (i.e., HDFS[5]) and

---

[5]http://hadoop.apache.org/

a high-level language for data analysis, i.e., Pig.[6] For example, a new stream of work [117, 234, 250] used Hadoop for large-scale graph storage and mining. They store and retrieve a large number of triples in a Hadoop file system.

1.3.3   Process Space

A BPMS is a centralized solution for supporting the definition, execution, and monitoring of structured business processes (those with explicit process representation). As pointed earlier, business processes in modern enterprises are followed in a dynamic, flexible, and sometimes ad hoc manner, and often a variety of information systems, resources, and services, not only BPMSs, are used to support their execution. The first step in understanding a business process is gathering information on its execution and a description of its underlying process model.

Understanding the process execution becomes more difficult as the number and the heterogeneity of involved IT systems increase. Another challenge is to obtain and correlate the relevant process events and artifacts to corresponding process executions. Moreover, processes are rarely well specified and explicitly modeled. Different people involved in a process may have different understanding of the scope and the details of the process and be interested in different aspects of it. In this context and from analytics point of view, a *process space* is an abstraction for understanding, representing, and managing various perspectives on the process execution in heterogeneous IT environments, specifically those environments in which processes are not necessarily supported by workflow systems, superimposed on the process-aware and process-unaware information systems that support the process execution.

Process space enables the representation of the process execution from various perspectives (different systems, business functions, or users) and at various levels of abstractions (detailed or abstract). Figure 1.3 illustrates the abstraction layers for understanding how process analytics can help in understanding business processes at the different phases of its life cycle in modern enterprises. In particular, Fig. 1.3 illustrates a layered architecture where at the lowest level, we have database entries. On top of this level, we have a representation of process-related entities (e.g., events and artifacts) and also a graph model that is superimposed over the entities. In order to understand available data (events, artifacts, data records in databases, etc.) in the context of process execution, we need to represent them as entities, understand their relationships, and enable the analysis of those relationships from the process execution perspective. In this context, graph models can be useful to represent all process-related data as entities and any relationships among them (e.g., event relationships in process logs with artifacts). In *process model* layer, we have the different ways that events can be grouped into process instances (different

---

[6]http://pig.apache.org/

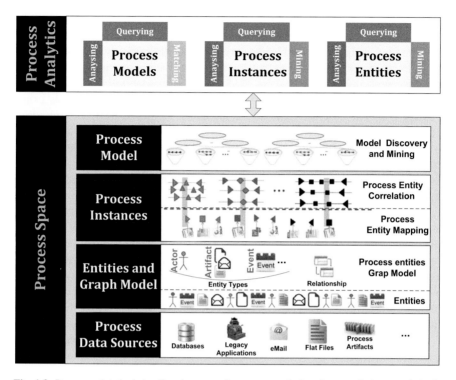

Fig. 1.3 Process-related abstraction layers and process analytics: how analytics can help in understanding business processes in modern enterprises

relationships over graphs are considered). In *process instance* layer, we apply model discovery/mining on each process instance set to get the process models. Finally in *process analytics* level, we have components for querying, process analytics operators, process model matching, and so on.

1.3.4 Business Process Analytics

Business analytics [38, 265] is the family of methods and tools that can be applied to process execution data in order to support decision-making in organizations by analyzing the behavior of completed processes (i.e., process controlling [348]), evaluating currently running process instances (i.e., business activity monitoring [7]), or predicting the behavior of process instances in the future (i.e., process intelligence [108]). In particular, the intent of process analytics can be motivated by performance, to shorten the reaction time of decision-makers to events that may affect changes in process performance, or compliance considerations, to establish the adherence of process execution with governing rules and regulations.

In enterprises, sources for process analytics data include activities, stakeholders, and business-related artifacts (and data) which are scattered across several systems and data sources. In particular, business process analytics might include events from multiple processes, data sources outside the organization, and events from non-process-centric information systems [265]. Existing works on business analytics focused more on exploration of new knowledge and investigative analysis using a broad range of analysis capabilities, including: (a) trend analysis, providing techniques to explore data and track business developments; (b) what-if analysis, in which scenarios with capabilities for reorganizing, reshaping, and recalculating data are of high interest; and (c) advanced analysis, providing techniques to uncover patterns in businesses and discover relationships among important elements in an organization's environment.

In order to apply analytics to business data, data sources (e.g., operational databases and documents) should be streamed into data warehouse servers (e.g., relational DBMS or MapReduce engine). This can be done using ETL (extract, transform, and load) or complex event-processing engines [303]. Multi-tier servers (e.g., OLAP servers, enterprise search engines, data mining engines, text analytic engines, and reporting servers) can be used on top of data warehouses for converting process execution data into knowledge and to support business users with decision-making process.

To understand the generated knowledge, a set of interactions between business users and expert business analytics is inevitable. For example, most analytic tools are designed for quantitative analysts, not the broader base of business users who need the output to be translated into language and visualizations (appropriate for business needs). A set of front-end applications (e.g., spreadsheets, dashboards, and querying approaches) can be used to address this challenge. Moreover, visual query interface and storytelling techniques [414] can be used to facilitate the understanding of business analytics results.

Several challenges have been introduced to characterize the gap between relevant analytics and the user's specific strategic business needs [108, 265] including the following: (a) *Cycle time* is the time needed for the overall cycle of collecting, analyzing, and acting on enterprise data. Business constraints may impose limits on reducing the overall cycle time. (b) *Analytic time and expertise* involve the time needed for analyzing generated knowledge from business data. Sometimes there are specific expertises necessary to analyze the result. (c) *Business goals and metrics* include various measurements such as cycle times, service-level variability, and even customer comments on a particular process that can be used to understand process analytics. For example, crowdsourcing, i.e., a process that involves outsourcing tasks to a distributed group of people [144], systems can be leveraged to understand the analytics results. (d) *Goals for data collection and transformations* are those that once metrics are identified, appropriate data must be collected and transformed into business data warehouses.

The ability for an organization to take all its capabilities and convert them into knowledge requires analyzing data about their customers and their suppliers. The wide adoption of CRM and supply chain management (SCM) techniques has allowed enterprises to fully interface and integrate their demand and supply chains. Trkman et al. [444] discussed the impact of business analytics on supply chain performance through investigating the relationship between analytical capabilities in the plan, source, make, and deliver areas of the supply chain. Moreover, online analytical processing (OLAP) techniques can be used for business reporting for sales, marketing, management reporting, budgeting and forecasting, and financial reporting.

In particular, OLAP servers can be used to expose the multidimensional view of business data to applications or users and enable the common business intelligence operations, e.g., filtering, aggregation, drill-down, and pivoting. In addition to traditional OLAP servers, newer "in-memory" business intelligence engines [382] are appearing that exploit today's large main-memory sizes to dramatically improve performance of multidimensional queries. For example, HYRISE [199] and HyPer [257] are both recent academic main-memory DBMSs for mixed (OLTP and BI) workloads. HyPer creates virtual memory snapshots by duplicating pages on demand when BI queries conflict with OLTP queries. HYRISE seems to be an offline analytical tool for deciding the proper grouping of columns and physical layout to optimize performance for a given mixed workload. On the commercial side, SAP Hana[7] has been recently presented as a unified and scalable database system for OLTP and OLAP using an in-memory column.

Recently, engines based on the MapReduce paradigm, which was originally built for analyzing Web documents and Web search query logs [131], are being targeted for enterprise analytics [108]. These approaches can be used to address challenges in real-time business analytics where the goal is to reduce the latency between when operational data is acquired and when analytics over that data is possible. Data platforms based on the MapReduce paradigm and its variants have attracted strong interest in the context of the "big data" [306] challenge: big data can be considered as a collection of datasets so large and complex that it becomes difficult to process using on-hand database management tools.

1.4   Goals, Structure, and Organization

This book will cover a large body of knowledge in process analysis including querying, analyzing, matching, and correlating process data and models to assist researchers as well as practitioners in understanding methods, tools, and techniques for modern process analytics. The goal of this book is to emphasize further the importance of process data analysis tasks. Throughout the chapters of this book, we will do a detailed study of the different approaches for analyzing business process at the different phases of its life cycle (Fig. 1.4).

---

[7]http://www.saphana.com/

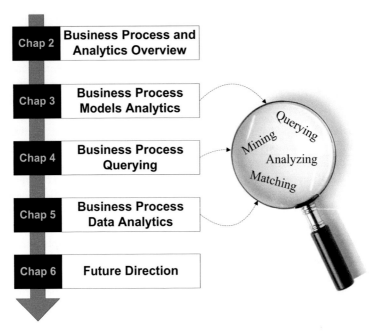

Fig. 1.4 Throughout the different chapters of this book, we will dive into detail with respect to the different approaches for analyzing business process

We provide concise and widely accepted definitions, taxonomies, and frameworks for BPM in Chap. 2. Chapter 3 presents process matching techniques that are useful in a series of process model analytics tasks. In Chap. 4, we focus on business process querying techniques and languages. In Chap. 5, we give an overview of different aspects of business data analysis techniques and approaches from process/dataspaces to data provenance and data-based querying techniques. Finally, Chap. 6 provides a summary and comparison of existing open-source and commercial software for process analytics, including real-world use case scenarios, followed by a discussion and future directions on some of the emerging and hot trends in the BPM area such as process spaces, big data for processes, crowdsourcing, social BPM, and process management on the cloud.

Chapter 2
Business Process Paradigms

There is no doubt that research and development in process management has significantly paved the path for a technological revolution. More importantly, process management technology is no less relevant today, as it was two or three decades ago at its inception. What has, however, mainly shifted is the primary *focus*. For example, in the 1990s, the focus was on *process automation* in order to reduce human involvement by enhanced systems integration and automated execution of the business logic. This is because total or partial automation of the process creates the unprecedented opportunity to gain visibility on process executions in the enterprise, whereas in the decade that followed, the focus had shifted into *process improvement*, which entailed understanding and analysis of business processes, implemented using various information systems and services and their integration. In this chapter, we provide an overview of the technological landscape surrounding business process management and set the stage for understanding the different aspects of analyzing business processes with the aim of improving them.

2.1 Introduction

More recently, the emergence of Web service technology has played a vital role in the implementation and integration of business processes. The main driving factors for their wide adoption is their success in reducing the costs and time needed to develop and integrate applications, in turn enabling to achieve greater degree of business flexibility [96]. However, in even more recent times, the success of Web service technology has by far transcended the initial expectation: in particular, Web 2.0 has provided a Web-scale sharing infrastructure—enabling unprecedented advances in social computing allowing real-time collaboration and communication among diverse participants. Similarly, the above has also provided a supportive platform for further advances, such as in cloud and crowd computing.

© Springer International Publishing Switzerland 2016
S.-M.-R. Beheshti et al., *Process Analytics*, DOI 10.1007/978-3-319-25037-3_2

However, as a result of the above, these advances have also influenced a new wave of process management support. *Social BPM* is one such example, which has sought to break down silos by encouraging a more collaborative, transparent approach to process improvement. Moreover, Social BPM takes advantage of social media tools (e.g., subscription feeds, social communities, real-time collaboration, tagging, walls, and wikis) to improve communication. A major benefit of social BPM is that it helps eliminate the barrier between BPM decision-makers and the users affected by their decisions [361]. However, while the Web (i.e., tools, services, apps) is well advancing, the formal and traditionally structured approach to *process management* is making it very difficult to embrace these advances. Web services today do not just provide functional or computational modules, they provide an entire *anything-as-a-service* stack, particularly *software as a service*, which is being used far and wide to support everyday tasks [46, 47]. In that sense, these software services (whether knowingly or unknowingly) are really being used to complete daily tasks as part of *unstructured processes*. However, the "process" often remains hidden and intangible. Recent recommendations have therefore aimed to set the focus about how to harness *process knowledge*, so that "unstructured" processes are visible and measurable, in much the same way as traditional BPM has provided for "structured" processes [362]. It is precisely these challenges and directives that should motivate the next generation of process management support.

In this chapter, we provide an overview of the technological landscape surrounding process management. We develop an advanced recognition of the potential gaps and thereby an appreciation for key areas of improvement needed to target successful future growth. Moreover, we utilize this knowledge in the remaining chapters to formulate a thorough understanding of how various process paradigms could be applied differently in process-based analytics. In Sect. 2.2, we thus begin by presenting an overview of the quintessential facets/dimensions often used to describe process types. These refer to, in general, (a) the *process paradigm* (i.e., the type of activities that are well supported),(b) the *implementation technologies*, and (c) the *representation model* or *language/s* available to a user, in order to describe this process. Accordingly, in Sect. 2.3, we examine the various identified implementation technologies (which are often independent from other facets). For this we have identified three main categories: (a) workflow engines, (b) rule engines, and more traditional (c) Program-coded-based solutions. In Sect. 2.4, we survey the relevant support tools categorized according to process paradigm: (a) structured, (b) unstructured/ad hoc, and (c) case management (semi-structured) processes, whereby, for each paradigm, we identify the various representation models/languages available. Moreover, this multi-faceted presentation proves indispensable in remaining for identifying how different process paradigms may exhibit different artifacts and/or entities that could be used as input into process-based analytics. Finally, in Sect. 2.5, we conclude with a summary and directions.

2.2   Dimensions for Characterizing Processes

There are many different characteristic terminologies to understand process-support systems and tools. For example, in the previous chapter, we referred to the notion of *structured*, *semi-structured*, and *unstructured* processes. This classification could more formally be referred to as the *process paradigm*, which in simpler terms represents the typical type (i.e., structure) of work activities that the process can handle.

However, while this may be the most general and commonly understood facet for describing processes and process-support system, there are several other dimensions that could be employed in order to characterize process systems. Most of which are arguably orthogonal and independent to each other. In fact, some of these terminologies while they may sound similar are actually quite different and thus tend to be a source of common confusion and misunderstanding. For example, process systems "implemented" using a *rule-based* engine should not be confused with process systems that offer a *rule-based* "language" for describing the process. The underlying "implementation technology" thus represents a significantly different dimension to the "representation model or language" that a given process system may offer. For example, Benatallah et al. [67] have proposed a model-driven approach (based on UML statecharts [216]) for defining composite-service processes which is then translated into rules for execution, while the system of Weber et al. [465] presented a rule-based language (specified via a spreadsheet interface) for defining composite-service processes which is then translated into WS-BPEL for execution.

In this section, we aim to demystify these concepts, thus providing a unified understanding of the various dimensions for characterizing processes and process-support systems and tools. Figure 2.1 illustrates the three main dimensions of processes that we have identified, as shown across the vertical axis. Across the horizontal axis, we show the different alternatives for each such dimension, which also represents the general conceptual and technological evolution. We further explain these various concepts in the following:

2.2.1   Process Paradigms

As mentioned above, "process paradigm" refers to the typical type of control structure the process or process system can handle. As mentioned above, there are generally known to be three such alternatives: *structured*, *semi-structured*, and *unstructured* [9, 73, 87, 130, 229, 362].

- **Structured processes** often mean that the functions to be executed, the sequence in which they take place, and the associated process control can be accurately described in advance. This category thus often includes operational processes that can be repeated (e.g., procurement or sales processes). Moreover, and in

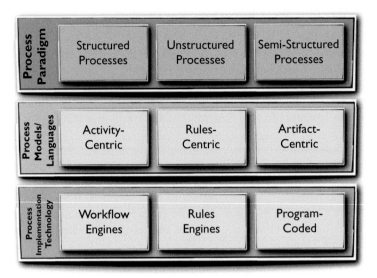

Fig. 2.1 Dimensions for characterizing processes

many cases, a structured process system may depend upon a predefined schema approach that often cannot be updated during execution.

- **Unstructured processes** on the other hand are generally very difficult to describe in advance. These types of processes may typically involve a significant degree of creativity, often having no fixed transaction description or process structure; for that reason a predefined schema-based approach thus clearly becomes unfeasible. Additionally, the activities in this type of process may be different in every case, plus they may change (or only become determinable) at some point during the execution of a process. It thus becomes difficult or near impossible to prespecify any structure over these activities—thus making the predefined schema-based approach clearly unfeasible.
- **Semi-structured (or case-based) processes** exhibit types of activities from both categories. While they may employ a certain structure, they still require a certain degree of flexibility comparable to that of unstructured processes. In particular, as its name suggests, part of a semi-structured process may be well defined, while other parts cannot be specified in full. This is as a result of either the type of functions to be executed is unknown, or, if known, the sequence in which they occur is unknown. However, this paradigm category can still be treated different to pure unstructured processes, as while these types of processes may not be entirely repeatable, there are often "recurring elements" or "patterns" that can be reused [130, 423]. For this reason, they are sometimes also referred to as "case-based" processes, as the overall process pattern might remain the same (as should be captured and possibly offered for reuse), while the specifics of the running process may nonetheless change on a case-to-case basis.

2.2.2   Process Representation Models/Languages

This dimension represents the language, model, or interface that a process designer may be offered in order to specify their desired process. We have identified three main categories of models/languages, as explained below. However, while not explicitly discussed, there could be considered a forth category: a visual interface. For example, Graupner et al. [190] presented an approach that provides a mind-map interface to drive business interactions. Other similar systems offered a spreadsheet-based interface [266] and form-based interface [465].

- The *activity-centric* approach represents the flow of control from activity to activity that is based on a specified sequence. Process tasks are therefore ordered based on the dependencies among them. Inherently this approach is therefore well suited for structured process. Whereas for activities that may be difficult to be planned in advance and thus require extra flexibility as well as ad hoc style processes, this approach would be considered too rigid to deal with in these cases [208, 240].
- The *rule-centric* approach unlike activity hierarchies is inherently less structured and less amenable to imposing an order in the flow. They are well suited to capture processes that have few constraints among activities and where a few rules can specify the process. However, although they provide this extra flexibility, a pure rule-based approach would only be suitable for small-scale processes with a relatively low number of rules (to be feasible to understand and maintain).
- The *artifact-centric* approach has activities that are defined in the context of process-related artifacts, which become available based on data events on the artifact. The intention here is to consider artifacts as first-class citizens and thus an abstraction to focus the process on core informational entities that are most significant (usually from an accountability perspective). The main features of this model are distinct life cycle state management and possibly also rules that may evaluate and constrain the operations of the artifact [79, 118].

2.2.3   Process Implementation Technologies

The underlying technology that is used to implement a process support could be considered as yet another dimension in characterizing process systems. As mentioned above, while the implementation technology behind some process system should not be confused with the application-layer aspects, such as the user language to interface the system, it has nonetheless been noted that the application-level and technology-level aspects need to support each other [415]. For instance, if a new application concept is developed that cannot be supported by existing implementation systems, it will not advance beyond the concept status and will thus not be capable of successful implementation [415].

- **Workflow engines** rely on the concept of a *business process* (or *workflow process*), which is typically specified by a directed graph that defines the order of execution among the nodes in the process. It also then introduces the concept of a *process instance*, which represents an execution of the workflow. A workflow instance may be instantiated several times, and moreover several instances of the same or different workflows may be concurrently running [29]. Furthermore, a central *workflow engine* is what executes workflow instances.
- **Rule engines** in general had sprouted from the need to manage and process large data streams, which also lead to complex event processors and subsequently reactive rule engine. However, while a pure rule-based orchestration scheme could be used for process execution, their applicability in nontrivial contexts has limited success, due to the number of rules required to describe a process. Nonetheless, in the context of process-execution systems, a rule engine could be applied as one means to help reduce the complexities of stand-alone workflow engines (in this case, it could be considered a somewhat partial hybrid solution). For instance, rules could be assigned to events, with each different occurrence of an event causing the process to take the appropriate route. Process descriptions can then be restricted to the more basic processing functions, with the rule engine ensuring that rules interpret events as they occur and activate the relevant process branch.
- **Programmed control flow** represents a third and rather more traditional alternative. In this case, we could think of the process being executed as a specification of hard-coded programmed statements. Currently (and after the advancement of workflow engines and rule engines), this method would be the least desired approach. However, we have included it for the purpose of completion—although, in some cases, this method might still be used partially, in cases where existing process systems are not capable to handle the required process task; in this case an application could be specially custom designed to handle these cases.

2.3   Process Implementation Technologies

2.3.1   Workflow Engines

As mentioned earlier, the key innovation introduced by workflow technology was the concept of "defining the business logic," as well as having a method for automating this "executable" process model. This means, in the same manner workflow systems were originally used to enable simple office automation by controlling how to dispatch information among "human" participants within an administration—it was soon realized that they could comparably be used for overall enterprise application integration.

Workflow engines could thus be seen as the *execution engine* for a defined business process. By enabling a developer to define and configure a business

Fig. 2.2 Workflow engine execution procedure (typically: scheduling and resource assignment)

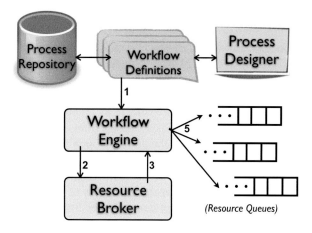

process, they are able to specify the order in which work activities should be executed, as well as to whom or what the work should be assigned [29]. More specifically, a "process instance" represents the actual execution of a process workflow. Such a process instance may be instantiated several times, and several instances of the same or different process may run concurrently.

A typical workflow engine basically acts like a scheduler; it schedules the work to be done and assigns it to an appropriate resource executor [29]. More specifically, a workflow engine functions as follows (as illustrated in Fig. 2.2): Whenever a new instantiation is created, the engine retrieves the corresponding process definition and determines the activities to be executed, with respect to their defined control flow. If an activity is to be executed, the engine then determines the resource executor to which this activity should be assigned for execution. (This could possibly be done by liaising with some resource broker that determines some user-defined resource selection policy.) The engine then places the activity to be executed into the activity queue of the selected resource executor. When ready to be executed, it retrieves an activity from the queue, executes it, and returns the result to the engine; the results are placed into the engine's inbound queue. Finally, routing between activities is performed by the engine simply evaluating the condition in order to determine which output arc should be activated and thus which activity should be executed next.

2.3.2 Rule Engines

The term "rule engine" generally refers to any system that uses "rules" in any form that can be applied to data to produce outcomes [113]. Rule-based systems sprouted with the increasing number of distributed applications that required processing continuous flows of data from geographically distributed sources at unprecedented rates to obtain timely responses to complex queries [124].

As a result, these requirements led to the development of a number of systems designed to process information as flows according to a set of pre-deployed processing rules. While some of these systems share many common goals, they also differ in a wide range of aspects, including architecture, data models, rule languages, and processing mechanisms. Nevertheless, after several years of research and development, two main models arose: the *data stream processing (DSP)* model [39] and the *complex event processing (CEP)* model [303]. DSP deals with the problem as processing streams of data coming from different sources to produce *new* data streams as an output; it is therefore commonly viewed as an evolution of traditional data processing supported by DBMSs. CEP, on the other hand, views information flowing as notifications of events happening, which have to be filtered and combined to understand what is happening in terms of higher-level events. Accordingly, the focus of this model is on detecting occurrences of particular *patterns* of (low-level) events that represent the higher-level events whose occurrence may require "actions": either *reactive* or *proactive*.

More recently, both these models have influenced modern-day event stream processors and rule-based engines; at the same time however, neither model alone is entirely self-sufficient; thus, many systems today adopt features from both worlds. Over the last decade there have therefore been various reaction-rule approaches developed—as they unequivocally play a vital role in making business, IT, and Internet infrastructures more agile and active—bringing in declarativity, fine-grain modularity, and higher-abstraction. In this section, we therefore present a brief description of pure reaction-rule approaches, together with examples of some existing systems. We have particularly positioned our discussion in the context of the *Web*—as the issues of notifying, detecting, and reacting upon events of interest begin to play an increasingly important role within business strategy on the Web and event-driven applications are being more widely deployed [75].

2.3.2.1 Event-Condition-Action Rules

Event-condition-action (ECA) rules confer the capability to *update data* found at local or remote (Web) resources, to *exchange information* about events (such as executed updates), as well as to *detect and react* not only to simple events but also to situations represented by a temporal combinations of events [75]. For example, such capability may be vital in e-commerce, for receiving and processing, buying, or reserving orders.

There are two possible strategies for communicating events on the Web: (a) the push strategy, which means a Web node informs other interested Web nodes about events and usually requires the interested parties *subscribing* for changes, and (b) the pull strategy, which means interested Web nodes have to specifically and periodically query (i.e., persistent polling) for data found at the other Web nodes in order to determine changes. While both strategies are useful, a push strategy is generally more advantageous over periodical polling, since it allows faster reaction, avoids unnecessary network traffic, and saves local resources.

Language Design/Semantics

An ECA rule, as its acronym suggests, have three main parts, applied in the form of: *ON Event IF Condition DO Action*, which specifies that we wish to execute the *Action* automatically when the *Event* happens, provided the *Condition* evaluates true. More specifically:

— **Event.** The event part drives the execution of the ECA rule specification, which makes this an important determinant for expressivity and ease of use of an ECA language or system. The general notion of an event could be thought as a *state change* of the system. However, what is observable (i.e., detectable as an event) largely depends on the boundaries and level of abstraction of the system. Typical examples of events could be messages from external or internal sources, the latter representative of system evens, or could also be time-related events. Events may also be *composite* representing a combination of atomic events satisfying some predefined pattern. The pattern is often called a composite event query, as it extracts data against the incoming stream of events, which could then be used in the condition or action part. Composite events may also contain temporal patterns, such as events before or after another event or date/time.

— **Condition.** The condition part usually expresses a further query of the data captured by the event patterns. As with event queries, condition queries also serve to determine whether or not the rule will fire (i.e., the action is executed). However, the condition query may often allow for more complex data patterns to be specified, including both lexical and temporal evaluators and comparators.

— **Action.** The action part of the rule usually serves the purpose of modifying the current system state; this may include updating persistent data as well as procedure calls and may in turn also cause new events to be raised. This is in contrast to the event and condition parts, which only detect that the system has entered a certain state without actually affecting the state. However, similar to the event part, we also often desire to execute more than one action—most commonly achieved by allowing several primitive actions to be specified in sequence. (Although, there are systems that enable more complex behavior, such as composite actions specified with one or more alternatives or a conditional execution plan.)

Existing ECA-Rule Systems

Aside from specific ECA systems, the *General Semantic Web ECA Framework* [63, 321] is an interesting research endeavor, as it proposes a general framework for reactive behavior, in that case in the context of the semantic Web. The generality here is given by the heterogeneity of the ECA rule components, which can be specified by using different event, condition, and action languages. The language used in writing an ECA component is given by means of the *ECA-ML* markup language, where ECA rule components are processed at Web nodes where a processor for the given language exists. Moreover, an ontology for reasoning and for easing the editing of ECA rules is also underway, which uses the *Resourceful Reactive Rules (r3)*[1] and *OWL-DL*[2] ontology as means for this.

More specific ECA rule systems include: *Prova*,[3] which is a combination of Java with Prolog-style rules, that employs ECA rules to enable distributed and agent programming. ECA rules react to incoming messages or proactively poll for state changes. Moreover, complex workflows can also be specified in the action part, as BPEL constructs are directly available in the language.

The *ruleCore* system[4] is another example and provides an engine for executing ECA rules, together with a GUI toolkit such as the *ruleCore Monitor*, which gives runtime status information on the engine. The ruleCore engine detects composite event patterns known as "situations," which includes sequences, conjunctions, disjunctions, and negation of events. Time-based and interval events can also be detected. Events are processed as XML documents. The action part of ECA rules can contain a number of action items, which specify that scripts are to be executed or events are to be generated.

Active XQuery [88] is another language proposed by extending the Web query language *XQuery* with ECA rules, which are adapted from *SQL-3* and thus referred to as triggers in this work. The event part of such a trigger specifies an affected XML fragment by means of an XPath expression and then an update operation (insert, delete, replace, or rename) on this fragment. The condition part is given by an XQuery WHERE clause. The actions available are the previously mentioned, simple update operations and external operations such as sending of messages.

2.3.2.2 Production Rules

Production rules were very popular since the 1980s, as a widely used technique to implement large expert systems for diverse domains such as troubleshooting in telecommunication networks or computer configuration systems. Classical production rule systems and most database implementations of production rules [418] typically have an operational or execution semantics defined for them [373]. Unlike ECA rule systems which introduce the concept of *events*, production rule systems introduce the concept of *working memory*. The working memory represents a finite set of data items (i.e., facts), against which the rules are executed. There is another data store, which is called the *production memory*, where the production rules itself are stored. The rule-execution process thus involves a *pattern-matching* process between the working memory and production memory. As illustrated in Fig. 2.3, the engine responsible for this pattern matching is often referred to as the *inference*

---

[1]http://centria.di.fct.unl.pt/~rewerse/wg-i5/r3/

[2]http://www.w3.org/TR/owl-guide/

[3]http://www.prova.ws

[4]http://www.rulecore.com/

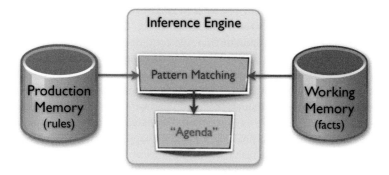

Fig. 2.3 Production rule-engine typical architecture and execution procedure

engine, which implements a discrimination network that is applied whenever a new fact is inserted (or updated) in the working memory. One common example is the *Rete*[5] algorithm [167]—although there are several optimized successor versions based on Rete, such as *TREAT* and *LEAPS*. Accordingly, if a match has been detected (between facts in working memory and rules in production memory), the *action* is fired for that rule. Moreover, it is interesting to note the actions themselves may change data, which in turn could match against other rules causing them to fire; this is referred to as *forward chaining* and in many ways acts akin to a database trigger, and we thus say this is a data-driven approach to reasoning.

Language Design/Semantics

Since production rules are executed based on a constant pattern-matching algorithm, the general semantic of the rule language has two main parts, in the form of: *WHEN Condition DO Action*. Meaning the engine matches data/facts against production rules to infer *Conclusions* which result in *Actions*. More specifically:

— **Condition.** The condition part of the rule expresses in which situation the rule should be elected for execution; it thus contains patterns describing the data that will trigger the rule. When evaluating a rule condition, the production rule engine will search the working memory for data that match all the patterns of the rule condition [75]. The expressivity of these patterns is thus dependent on the data model of the rule language (although typical examples are constraints on the class of objects or value of attributes). Constraints that involve a single data item from the working memory are called discrimination tests; constraints that involve several data items from the working memory are called join tests [75].

---

[5]The latin word "rete" means "net" or "network".

— *Action.* The action part describes what should be performed as part of executing
the rule. Whenever an appropriate fact (or set of facts) from the working
memory matches all the patterns of a rule condition, a *rule instance* is created.
Executing a rule instance thus means interpreting the statements in the rule
action and firing the action. Production rule languages often rely on a foreign
data model borrowed from a programming language, in that case, we expect the
statements that can be found in the action part of a production rule usually are
those that can be found in any procedural language: assignments, conditionals,
and loops. For instance, a production rule language using the Java object model
is likely to express the action parts of its rules in Java or a Java-like scripting
language. If the production rule language matches XML documents, specific
statements would have to be introduced to express the action parts of those
types of rules.

Stateless Versus Stateful Semantics

An important distinction with respect to production rule systems is the concept
of *statefulness*: the *stateless* case corresponds to the inference engine evaluating
rules in stateless session; this means it does not utilize any inference. This can
sometimes be referred to filtering applications, where the rules are used to scan
a flow of objects on a one-by-one, or tuple-by-tuple, basis. Common use cases
for this may include validation (e.g., is this person eligible for a mortgage?),
simple computation/calculation (e.g., compute a mortgage premium), and routing
and filtering (e.g., filter incoming messages, such as emails, into folders, or
send incoming messages to a destination). The *stateful* case, on the other hand,
corresponds to applications that correlate data items or that infer information from
the existing data items. Stateful sessions are thus longer lived and allow iterative
changes over time. In these applications, and in contrast with the stateless case
described below, the action of one rule may heavily influence the eligibility of other
rules. As a consequence, the rule engine must carefully take update notifications into
account in order to ensure that the truth value of the rule conditions, and thus the list
of eligible rules, is known at any time [75]. Common use cases include: monitoring
(e.g., stock market), diagnostics, fault finding, compliance, validation, etc.

Existing Production Rule Systems

The following are some popular production rule systems that we have identified:

The *OPS5* (Official Production System) language, designed in the 1970s by
Charles Forgy, was one of the first production rule language to be used in expert
systems. OPS5 uses a forward-chaining inference engine; rule programs are thus
executed by scanning the "working memory." The system although ancient is still
arguably valuable and claims to be efficient for high-scale problems involving
hundreds or thousands of rules.

Jess[6] is a production rule language for the Java platform, which implements an advanced Rete algorithm. Rules may be specified using the Jess rule language, which could be written as both a stand-alone program and via the Java API (after embedding the Jess Java library). Alternatively, rule-based logic could also be specified in XML, which is particularly useful if embedding in Web-based application. The engine provides both forward chaining (i.e., reactive and data driven) and backward chaining (i.e., derivation, passive queries). Interestingly, it also provides the `defquery` construct which has no rule body but is used to search the fact knowledge base under direct program control. This query returns the list of all fact tuples of the working memory matching the rule condition.

Blaze Advisor [85] provides a number differentiating features in comparison to some of the more standard rule system (like those mentioned thus far). In particular, it provides a "reuse" capabilities over the rule repository; this means developers are enabled to work in a coordinated manner as teams and leverage each other's work by sharing and reusing rules, rule sets, rule flows, and object models. The rules in the repository could be curated as an XML file, a database, or even an LDAP directory. With respect to the rule language, it also offers a simple English-like Structured Rule Language (SRL) for writing rules. Furthermore, it includes a visual development environment for writing, editing, and testing business rule services that are executed using a sophisticated rule server. At the same time sophisticated rule-set metaphors (decision tables, decision trees, and scorecards) provide a way for nonprogrammers to author rules as well.

PegaRules [376] introduces different types of rules: declarative rules (computing values or enforcing constraints), integration rules (interfacing different systems and applications), transformation rules (data), as well as decision-tree rules. Similar to *Jess*, its execution environment provides both backward and forward chaining capabilities, in order to determine known and unknown dependent facts. Moreover, it provides an HTML form to configure and manage rules.

2.3.3 Program Coded

In more traditional application systems, the approach for a process-driven solution was often treated akin to the implementation of any other application. This meant that the sequence of individual statements simply had to be hard-coded in the computer program. Accordingly, the process designer (in this case, an experienced software developer) must specify the tasks which are then executed in the order defined by the program system. Inevitably in this approach, the available option for executing individual transactions is very limited. Over the course of time, new layers of abstractions were added: such advances included the introduction of *workflow engines* and *rule engine*. To put things into a technological perspective,

---

[6]http://www.jessrules.com/jess/docs/

we outline below: (a) how these technologies were built upon the underlying computer programming layer, (b) how they compare/intersect with each other, and (c) moreover, what important improvements were achieved.

Workflow engines were in fact designed with inherent similarities to programs written in general-purpose programming languages, such that they are characterized by the invocation of several functions in an order specified by some flow of logic. The effects of loops and condition statements are modeled by inserting appropriate routing nodes. In addition, and similar to programming languages, a workflow could typically have variables (data items) that can be passed as input or taken as output from work activity invocations. These variables could be used to evaluate routing conditions and to pass data among activities [29]. However, despite these similarities, workflow systems present significant differences in contrast to traditional programming techniques. Foremost, it introduced a higher level of abstractions that appealed to nontechnical domain experts—thus reducing the dependence on experienced programmers. Moreover from a more technical perspective, it provided high scalability. While procedures invoked in the context of computer programs are generally short-lived, workflows typically made it possible to compose coarse-grained activities and applications that can last hours or days. It also provided improvements to the granularity at which composition takes place. Workflow systems allow the composition of large software modules, typically the entire applications, thus embodying the notion of "mega-programming" [473]—as a shift in software engineering toward programming as an exercise in composition of large modules, rather than programming from scratch.

Similarly, compared with general-purpose programming languages and frameworks, *rule engines* brought in declarativity, fine-grained modularity, and higher abstractions [75]. Working together with workflow engines, rule engines provided an additional step ahead, as it allowed splitting out the process control logic, otherwise embedded in workflow systems, and transferring it to a separate rule control system [415]. Since essentially both workflow systems, alike a traditional hard-coded application, require a top-down approach, support for flexibility is limited, and making changes can prove expensive in terms of time and money when restarting from scratch [29]. Rules on the other hand can be individually described in business terms and are also easy to change, thus promoting flexible handling of organizational changes [415]. Moreover, modern rule-based frameworks enable natural-language-like syntax, as well as support for the life cycle of rules. These features make it easier to write, understand, and maintain rule-based applications and also appeal to nontechnical users.

2.4 Process Paradigms: A Survey of Frameworks and Tools

2.4.1 Structured Processes

2.4.1.1 Activity Centric

Activity-centric, *structured-process* systems usually refer to workflow management systems (cf. Sect. 2.3.1) or more recently complete business process management systems. This is because in essence, the notion of a "business or workflow *process model*" precisely fits the definition of the "activity-centric" model, in that it represents a directed graph (called a *flow graph*) that defines the order of execution among the nodes in the process [29]. Likewise, the notion of "business or workflow *process instance*" precisely satisfies the "structured" characteristics of these types of process—as usually a process instance refers to an unchanging instantiation of the process model, which can only be suitable when the process is well structured and without the need for ongoing flexibility.

There are in fact several activity-centric languages used to define structured processes, such as: WS-BPEL [161] (or its predecessors XLANG [329] and BPML [440]), XPDL,[7] YAWL [449], BPMN [365], etc. Albeit more generically, there has often been identified the following set of dimensions used to characterize and pose requirements on these various languages (we discuss below, using BPEL as an example):

1. **Component model** defines the nature of the elements in the process that can be composed, in terms of the assumption the language makes on these components. At one extreme, a model could assume that components implement a specific type of Web service standards, such as HTTP, SOAP [337], WADL, etc., while, at the other extreme, it may only make very basic assumptions, such as it assumes that the components only interact by exchanging XML. The advantage of the former is easier syntax at the cost of limiting heterogeneity, while the advantage of the latter means more flexibility although at the cost of more complicated syntax.

   For example, BPEL has a fine-grained component model, consisting of *activities*, which can be *basic* or *structured*. Structured activities, as in the activity hierarchy approach, are used for defining the process orchestration, while basic activities represent the actual components, which correspond to the invocation of WSDL operations performed by some Web service. This manifests itself in the *invoke* activity (representing a request-reply or a one-way operation of a service), *receive* activity (representing the receipt of a message), and *reply* activity (representing a message sent by the process in response to an operation invoked). Other types of activity may serve data-manipulation functions, such as *assign*

---

[7] http://www.xpdl.org/

or *copy* activity, or the *wait* activity, to define points in the process where the execution should block for a certain period of time or until a data/time is reached.

2. **Orchestration model** defines how different elements of the process model work together in order to form a coherent whole. In particular, it may specify the *order* in which activities are executed, as well as any conditions under which this may need to be considered.

   For example, BPEL allows the definition of structured activities, which can group a set of other structured or basic activities, to define ordering constraints among them. More specifically, the following structured activities can be specified: *Sequence*, which contains a set of activities to be executed sequentially; *Switch*, which includes a set of activities, each associated with a condition; *Pick*, which includes a set of events (such as the receipt of a message or the expiration of a time alarm), each associated with an activity; *While*, which includes exactly one (basic or structured) activity, which is executed repeatedly, while a specified condition holds true; and *Flow*, which groups a set of activities to be executed in parallel.

3. **Data and data-access model** defines how data is specified and how it is exchanged between components. Data itself can usually be divided between *application-specific* data (e.g., the parameters sent or received as part of message exchanges) and *control-flow* data (e.g., data relevant to evaluate branching conditions). The second aspect to this is the data-transfer aspects, which deals with how data is passed from one activity to the next.

   For example, BPEL maintains the state of the process and manipulates control data by means of a variable. In this case, variables are analogous to variables in a conventional programming language. They are characterized by name and type, specified as a reference to a WSDL message type, XML schema simple types, or XML scheme elements. Variables can thereby be used as input or output parameters to service invocations as well as be updated through an expression defined on the variable itself. (This could be useful, e.g., to increment counters.) Moreover, the value of variables can also be passed between activities via an intermediate *assign* activity—to enable this an XML query language such as XPath,[8] would normally be useful.

4. **Exception handling syntax** defines how exceptional situations (i.e., deviations from the expected or desired execution) that occur can be handled, without resulting in the process being aborted. Exceptions can typically be caused by failure in the system, or they can be situations that although contemplated are infrequent or undesired. (*Transactions*, as we will see next, are one possible way to handle exceptions.)

   For example, BPEL essentially follows a *try-catch* approach for exception handling. This is typical in structured orchestration model process systems, where activities can be nested into each other. Basic or structured activities implicitly define a scope, or it can be explicitly defined, whereby scope-defining

[8]http://www.w3.org/TR/xpath

elements can include the specification of one or more *fault handlers* that is characterized by a *catch* element which defines the fault it manages. Faults can either be generated during the execution of an activity within the scope or explicitly thrown within the process orchestration schema, via the *throw* activity. When a fault is detected, the BPEL engine terminates the running instances in the scope and executes the activities specified in the fault handler for that scope.

5. **Transactional syntax** defines which transactional semantics can be associated to the process, and how this is done. While traditional approaches to this involves the definition of *atomic regions* exhibiting the all-or-nothing property, recent solutions have adopted the *compensation* approach, where committed executions are semantically undone by executing other operations. In some cases, the compensation may be transparently handled by the engine, or else the process designer may want to explicitly define the operations to perform compensation.

   For example, BPEL combines exception handling approaches with transactional techniques. Thus, for each scope, it is possible to define the logic required to semantically "undo" the execution of activities in that scope. The compensation logic is specified by a *compensation handler*, which takes care of performing whatever actions are needed to compensate the execution. Its invocation can either be explicitly initiated by a *compensate* activity or occur automatically as part of the default fault handler.

6. **Correlation syntax** is important where there are multiple instances of the same process running. Correlation thus allows to initialize a business process transaction, temporarily suspends activities, and then recognizes once again the correct instance when that transaction resumes. Correlation is usually realized, by *correlation sets*, which are initialized with values from process inbound or outbound messages, as well as *correlation properties* which contain a *name* and a *data-type* as well as define the correlation sets with which they are associated. In this manner, we could also think of correlation as serving to map received messages to the correct process instance.

   For example, BPEL includes instance routing syntax to cater for those cases where routing is not transparently supported by the implementing engine. It addresses this by defining how to correlate messages with instances, based on message data. In this case, a correlation set essentially identifies a set of data items that can be associated with messages sent or received, within the invoke, reply, or receive activities. In this way, by associating a pair of messages to the same correlation set, it thereby means that if these messages have the same value for the correlation set, they belong to the same instance. Moreover, as the characteristics that enable the identification of an instance from the message data may change, BPEL also enables multiple correlation set to be defined, each of which may be relevant for different stages of the process.

As examples of process systems that fit within this category, we have examined a selection of such activity-centric structured-process systems, which we briefly present below in the following:

Enhydra Shark[9] is an extensible and embeddable Java workflow engine framework completely based on WfMC specifications [227]. Shark (for short) can be used as a simple Java library in a Servlet, a swing application, or in a J2EE container. As for front-end features, it provides by default, a graphical editor called *Together Workflow Editor (TWE)*, which supports the design of the organizational perspective and the workflow language used is XPDL. However, the platform does not offer a Web-based environment. The system is based upon a rather traditional middleware platform (CORBA), although it provides support for integration with most database management systems and offers mechanisms that support the exceptions treatment during a process execution.

JawFlow[10] was developed by *Vincenzo Marchese* and is a workflow engine written in Java. It can be customized using activities written in Java or in any scripting language supported by the *Bean Scripting Framework*. This workflow engine does not provide a process editor, such that any editor supporting XPDL can be used, although it does provide a Web-based process administration environment. This system is based upon a middleware platform (Java RMI and CORBA) and can be integrated with any database management system. It also offers mechanisms that support error handling during the execution of a workflow process.

JOpera[11] is built as a collection of plug-ins for *Eclipse*. It primarily acts as a service composition tool that offers a visual language and an execution platform for building workflow and business processes. It includes a graphical modeling environment, a lightweight execution engine, and also a set of powerful debugging tools in order to support the natively iterative nature of service composition. JOpera specifies its own proprietary workflow language, called the JOpera visual composition language. As its front end, the environment offered is based on the Eclipse workbench, although it is not immediately practical and arguably not a very user-friendly management environment. However, it provides integration support for the most popular DBMS and supports a simple exception-handling model.

WFMOpen[12] is a Java/J2EE-based engine that exposes a set of Java interfaces that defines API for business process management. It is thus extensible and could be used as the core for any process-based application implementation. The system is based upon a middleware platform (Java RMI, CORBA, and SOAP). It used XPDL with some extensions to define workflow processes as well as enable the Java Workflow Editor (JPEd) to be used. It also provides a Web-based management environment, although it has limited DBMS integration support, as integration can only be achieved using the default DBMS of the system. However, it offers built-in solutions for handling exceptions during a process execution.

---

[9]http://sourceforge.net/projects/sharkwf/

[10]http://jawflow.sourceforge.net/

[11]http://www.jopera.org/

[12]http://wfmopen.sourceforge.net/

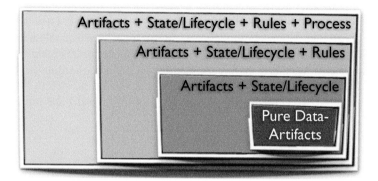

Fig. 2.4 Artifact-centric conceptual and technological landscape

2.4.1.2 Data/Artifact Centric

In the context of processes, *data* or *artifacts* commonly refer to some form of tangible, digital product [264] that is either consumed or produced as part of the process enactment [332]. Artifacts may have attributes, be aggregated, include relationships to other artifacts, or be versioned [264]. The artifact-centric approach thus focuses on defining the behavior of a process in terms of "what" type of work or actions are allowed but not "how" work occurs. This approach is different to the process-centric approach, which only focuses on "how" things happen. Process descriptions thus often define the process areas, activities, and tasks that must be performed to get the job done. Upon analysis, we consider the artifact-centric conceptual and technological landscape to consist of four layers of concentricity, as illustrated in Fig. 2.4.

At the core, we have *pure data artifacts*, which is only concerned with the definition and management of the data aspect of artifacts; at the second layer we have *artifacts combined with state-based life cycles*, where apart from the data model, a life cycle model based on states and transitions is defined; at the third-layer, we further combine *event-driven rules* as a means for provided automated support of life cycle transitions; finally at the fourth, outer layer, we integrate all the above with aspects of "process" technology, often resulting in an *actionable* framework that can be mapped or linked with a process/workflow engine.

Pure Data-Artifacts

As the starting point, a pure data/artifact based solution is no different to "document management" techniques based on work developed by the document engineering community. In this area, models and tools are developed around the concept of documents, which essentially represents a particular type of resource. In this category, the type of support may include: data/artifact management tools, such as structured data repositories [186], as well as the notion of document-centered collaboration, incorporating real-time and concurrent authoring. Nonetheless, in all these cases, the notion of "process" is almost nonexistent (or rather managed invisibly), and in fact this category of support is best suited for unstructured and ad hoc environments. We therefore limit further discussion in this section, as while we mention here for the sake of completion, the reader may be directed to Sect. 2.4.2 on "Unstructured (Ad Hoc) Processes" support, for further details and discussion.

Artifacts + State/Life Cycle

As an added level of support, artifacts may also define a "life cycle," akin to a directed graph of *states* and *transitions*. In this manner, *states* represent key values of the artifact, and *transitions* represent changes to the artifact which move it from one state to another. One benefit of the artifact-centric approach in this case is that it provides a single place to constrain actions or decisions which move an artifact from one state to another, without needing to specify specific actions that induce that move [264]. An artifact-centric model thus enables to focus on the artifact itself as a first-class citizen, together with their transitions—while leaving the specific process or task sequence which causes the transition up to the individual. At this level of support, we assume that the life cycle of the artifact is driven manually by human workers.

IBM Governor [476], for example, is a tool that adopts the artifact-centric model for specifying and managing *governance* processes over software development projects. Governance of software development can be characterized as an iterative process concerned with the *goals* and how they are expressed in the *decisions* made by different *roles* involved with the project. In this case, "artifacts" represent any digital products of the software development process, such as: code, models, tests, change requests, requirements, documents, project plans, etc. Thereby, "states" represent the key stages/goals that a given artifact may or must acquire during its lifetime, while "transitions" represent the decisions to be made by one ore more stakeholders (holding some role), before the artifact can move to a different state. In many cases such as this, the most natural characterization of change to an artifact is the making of a binary decision, corresponding to a guard from the source to target state (e.g., *Decision=Yes*; *Decision=No*), as illustrated in Fig. 2.5a. The decisions are performed by human workers; however, *Governor* extends the basic artifact model by controlling not only *what* artifact states and transitions are allowed but also *who* is allowed or required to perform them. To do this, an RACI matrix[13] is also specified with decisions down the rows and roles across the columns, as illustrated in Fig. 2.5b. Each cell entry then represents the rights and responsibilities.

---

[13]Further described by *Hallows et al.* [211], RACI is an acronym for the possible types of rights and responsibilities, namely, "**R**esponsible", "**A**ccountable", "**C**onsulted", and "**I**nformed"—where each have specific semantics.

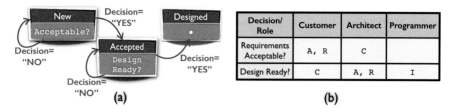

Decision/Role	Customer	Architect	Programmer
Requirements Acceptable?	A, R	C	
Design Ready?	C	A, R	I

(a)   (b)

Fig. 2.5 (**a**) Requirement life cycle for software development artifacts; (**b**) RACI matrix for specification of roles and Responsibilities

Gelee [40] is another example in this category that provides life cycle management of Web-based resources. Artifacts are thus any URI-identifiable and accessible object/resource. The work also provides a graphical tool for modeling artifacts based on finite-state-machine concepts; albeit, the contributions are not so much in the model but rather in the instantiation and execution model, as well as in the light binding between models and instances and dealing with the heterogeneity of possible resources. States in this case correspond to phases that the resource can undergo; and these states (or phases) can also have associated actions. However, the actions in this case do not automate the transition of life cycle phases; and the engine is simply the human life cycle owner. Rather actions simply provide a way of automating the possible operations on a given resource (e.g., modifying content or access rights), which is automatically executed upon entering a phase.

Artifacts + State/Life Cycle + Rules

As a second level of support, the artifacts and life cycle model introduce event-driven rules as a way to enable automation within the life cycle flow. In these cases, it is common that the modeling is achieved using state machines. As before, the semantics remain the same, in that *states* represent key values of the artifact, and *transitions* represent changes to the artifact which move it from one state to another. However, automation can be enabled by adopting state-machine concepts where transition also incorporate the notion of: an *event*, a *guard*, and an *action*. Guards are used to make sure that the transition only can trigger if the evaluation of the guard is true. Therefore, it becomes possible to create transitions with the same event and different guards. Depending on the guard different actions can also be executed or target states reached. Furthermore, *actions* can also be associated with states where there are three kinds of *state* actions that are available: (a) the *entry action* is executed when the state is activated; (b) the *do action* is executed after finishing the entry action and anytime while in that state; and(c) the *exit action* is executed when the state is deactivated. This approach is often seen used in industries that require to standardize their data objects with life cycles in order to facilitate interoperability between industry partners and enforce legal regulations [405].

Artifacts + State/Life Cycle + Rules + Process

Finally, the notion of a system combining together the artifact-centric approach with its life cycle (state and transitions) model, as well as event-based rule capabilities, plus the integration of process aspects into a holistic unit, could be referred to as "artifact-centric business process modeling/management (AcBPM)" or for short "business artifacts (BA)" [80, 274, 300]. The underlying motivation for this emerging technology is the aspiration toward a flexible solution for otherwise rigid BPM systems. Traditionally, most BPM frameworks (e.g., [290]) have used meta-models oriented on activity flows, with the data manipulated by these processes considered as second-class citizens. Alternatively, other frameworks (e.g., [181]) focus on the documents that track the business operations, with the process meta-model typically impoverished. For both, associated requirements, business rules, and business intelligence are based on conceptual meta-models only loosely connected to the base model. This disparity adds substantial conceptual complexity to models of business operations and processes, making them hard to understand [118]. The *AcBPM* or *BA* approach thus aims to strike a hybrid between *data aspects* and *process aspects*. It particularly focuses on describing the data of business processes, by characterizing business-relevant data objects and their life cycles and linking this with related process tasks. It is claimed that this approach fosters the automation of business operations and supports the flexibility of workflow enactment and evolution [80, 231, 232]. It also claims to facilitate effective collaboration of the business processes as the life cycle of an artifact may span multiple business units when it evolves through the process.

Cohn et al. [118], for example, defines "business artifacts" to include both an *information model* for data about the business objects during their lifetime and a *life cycle model*, describing the possible ways and timings that tasks can be invoked on these objects. (The work is based on research headed by *IBM Research Labs* and presents in various versions covering different aspects: *Nigam et al.* [358] can be cited as introducing the concept of business artifacts and information-centric processing of artifact life cycles; accordingly, *Bhattacharya et al.* [78, 80, 300] present further studies on artifact-centric business processes.) The work is also referred to as: business entities with life cycles (BELs). The information model is described to have different slots for the information gathered during the life cycle of the artifact; while it starts out largely empty, attributes are filled in over time. Moreover, an artifact's information model allows to "cluster" the various kinds of data which correspond to the stages in the business entity's life cycle—rather than breaking into separate entities as is common in typical databases. The benefit of this approach is to enable better communication between stakeholders, as it facilitates better understanding/vocabulary of data. The life cycle model thereby defines the key business-relevant stages through which the artifact may pass, with transition edges corresponding to tasks that can be performed by participants. These tasks relate to actions/operations that can be invoked as part of a "process." Therefore, a key feature of AcBPM is to link the artifact model with these types of process tasks. ECA-like rules are often used for this.

Hull et al. [233] proposed an evolution of this work, called *business entities with guard-stage-milestone life cycles*, abbreviated as *BEL[GSM]* or simply *GSM*, in which the life cycles for BELs are much more declarative than previous work. The work also claims an actionable framework, where the artifact model can be automatically loaded onto a workflow engine to create a deployed system; the following experimental implementations are provided: *BELA's FastPath* tool [427] and *Siena prototype* [119]. It has been designed to: (a) include constructs that match closely with how business-level stakeholders conceptualize their operations; (b) enable the specification of those constructs in a precise, largely declarative way; and (c) enable a relatively direct mapping from a *GSM business operations model (BOM)* into a running implementation. In this manner, the work had posed an interesting parallel with case management. This in fact was the precursor that ultimately merged the artifact-centric and case-centric approaches into a unique effort, as testified by the OMG standard called CMMN [366] (albeit currently in its beta version).

Redding et al. [393] and ***Küster et al. [278]*** are other examples for providing techniques to generate business processes that are compliant with predefined artifact life cycles. Similarly, event-driven process modeling, such as *event-driven process chains (EPC)* [326], also describes artifact life cycles integrated by events.

2.4.1.3 Rule Centric

In the context of rule-based processes, it seems appropriate to precede our discussion with a mention about the *business rule approach (BRA)*; as in many ways this discipline represents the *formalism* of the synergy between *rule-based programming* (including pure rule-based processes) and *business processes* (in the context of BPM).

Business Rule Approach (BRA)

With respect to BRA, a "business rule" (BR) could be considered as a statement that expresses (certain parts of) a business policy, defining business terms and defining or constraining the operations of a "process," in a *declarative* manner [458]. Business rules can be enforced on the business from the outside environment by regulations or laws, or they can be defined within the business to achieve the goals of the business. For example, a business policy in a car rental company is "only cars in legal, roadworthy condition can be rented to customers" [220]. Business rules are declarative statements: they describe *what* has to be done or hold, but not *how*. In the current literature, *four* main types of business rules have been defined:

1. **Integrity rules** constrain the flow and integrity of a process model. Given that we may consider a process model to consist of dynamic entities which change their state during process enactment, integrity rules express constrains over such evolving states and state transition histories. There are a few subtypes of integrity rule: (a) *State-transition constraint* controls the transition of a process from one state to another. State constraints must hold at any point in time (e.g., "a customer of the car rental company Discount CarRental must be at least 26 years old"). (b) *Dynamic integrity constraint* controls the state model on the admissible transitions from one state of the system to another (e.g., a rental order is defined by the following transition path: "reserved → allocated → rented → returned").
2. **Derivation rules** are where elements of knowledge is derived from other knowledge. Derivation rules capture terminological and heuristic domain knowledge that need not to be stored explicitly because it can be derived from existing or other derived information on demand. For example, "the rate of rental is inferred from the group of the car."
3. **Reaction rules** specify actions to be taken in response to occurrences or nonoccurrences of facts. They thus state the condition under which the action must be taken, as well as the actual action to invoke. (Reactions are precisely either *ECA* or *production* rules, as we presented earlier in Sect. 2.3.2.) As an example, a reaction rule might be "WHEN customer request received DO background check of customer."
4. **Deontic rules** express the power, rights, and duty of an agent for performing or coordinating work and thus define the deontic structure of an organization, guiding and constraining the actions of agents. General examples of this may be: (a) *authorization*, where if commitments are assigned to agents to perform work, work assignments depend upon the properties of the work, agent, or state model, whereby rules constrain the domain in which the assignment can range over, and (b) *event subscription constraint*, which represents another form of access control which constrains the agents to receive events in the process. A more specific example of a deontic assignment statement could be: "only the branch manager has the right to grant special discounts to customers."

In general therefore, the BRA may be considered as another *tier* upon the traditional business process management systems, although as a subset, it could also refer to a *purely* rule-based approach (e.g., we illustrate below two such examples). In any case, the current idea is to therefore effectively "integrate" business rules within existing BPMS systems, including its modeling languages, with the aim of enhancing business agility. In latter part of this section, we therefore examine: some of the work leading to merging business rules and business process management systems, as well as some examples of these partial-hybrid/dual-support process systems that have been proposed. Although, we first examine two examples of pure rule-based systems.

JBoss Drools Expert[14] is a Java open-source (Apache 2.0 License) business rule engine that implements an advanced version of the *RETE* and *LEAPS* pattern-matching algorithms. Since facts typically represent Java objects, the discrimination network is formed by nodes such as: *OpbjectTypeNode*s, which filter based on the class of object, and *AlphaNode*s, such as *name == "cheddar"*, as well as input-type nodes (also known as beta-type nodes) such as *JoinNode*s or *NotNode*s, which are used to compare two objects, and their fields, to each other. The objects may be the same or different types, such as: *Person.favouriteCheese == Cheese.name*. The system implements both stateless and stateful knowledge sessions and, moreover, integrates a conflict resolution agenda based on priority as well as the LIFO principle. The rule language mostly extends the basic Java constructs (particular for the action part); however, there are several additional language constructs added that make up the *Drools rule language, "DRL"* Such as the `insert` statement, which is particularly useful to enable an inference relationship.

In addition, Drools Expert provides a rule "template" feature, which is useful in the cases where a group of rules is following the same arrangement of patterns, constraints, and actions (i.e., differing only in constants or names for objects or fields). This means, if a rule template file is written containing the textual skeleton of a rule, the provided template compiled would then flesh out the actual values for that rule for their instantiation. Moreover, Drools Expert makes use of this feature to enable a spreadsheet-based (therefore, graphical, familiar, and computational) environment for declaring rules. A dedicated compiler thus serves to analyze the spreadsheet and translates it into rules.

The Demaq Platform [86] presents a rule engine specialized for processing XML messages. The motivation stems from the rapidly growing number of active distributed systems over the Web that often participate in asynchronous communication via XML messaging. Referring to this as "ActiveWeb" nodes, these includes: event-notification systems using RSS/Atom feeds, business Web services, and even new end-user interface architectures such as AJAX. Moreover, this work adopts an alternative implementation approach (compared to conventional rule systems). Instead of having both incoming and outgoing messages travel through all layers of the implementation stack (including middleware and network layers), they proposes a fully declarative, executable rule language for the specification and implementation of ActiveWeb nodes. As a result, the claim here is better productivity enabled by the opportunity to move the responsibility for implementation details from the programmer to the processing system. Declarativity also facilitates data independence, which in the case of message processing means that aspects such as message persistence and recovery, message retention, or transport protocols are transparent to the programmer unless their control is explicitly desired. Moreover, Demaq also extends the basic rule language with a "queue definition language." *Queues* provide physical message containers that decouple message insertion from processing. In turn they also introduce the notion of *slices* (a virtual

---

[14]http://docs.jboss.org/drools/release/5.5.0.Final/drools-expert-docs/html_single/

queue), which can be used for creating logical groups of related messages. Queues and slices pose interesting capabilities, as they could potentially be used to manage correlation/segregation of messages and thus also support the notion of process instances.

Synergies Between Business Rules and BPM Systems

When it comes to specifying processes, although stand-alone business process management systems may enable nontechnical domain experts (e.g., business workers) to define and manage their process using high-level modeling notations (e.g., via BPMN [365]), specifying business rules in the same way is often more difficult.

In fact, one of the main challenges of BPM is to ensure the compliance between the modeled processes and the real-world processes [422]. While a process can be modeled at the high level, in most cases there is a need for detailed, formal specification of such process showing its decomposition and relationships with other elements of the whole business model. Moreover, a detailed business process specification often requires the process designer to work with *both* the *static* and *dynamic* aspects of the process. The aspects related to modeling of activities (tasks), control flows, decision points, and other artifacts represent the dynamic (i.e., what BPMN alone may offer). At least one of the provisions in supporting flexibility for the dynamic aspects of process modeling can be attributed to the long-standing line of research in constraint-based declarative process modeling. For example, *Declare* [377] provides multiple declarative languages (*ConDec* [378] and *DecSerFlow* [448]) that aims to avoid process change (i.e., increasing agility) by preventing users from over-specifying. Thus, instead of implicitly defining the order of activities, it is possible that an imperative approach can be replaced by a declarative one. Meaning, any order that does not violate the constraint is effectively allowed.

However, static aspect cannot be left aside either as this is where the main business entities, their structure, relationships, and constraints are specified, and that is what forms the core of a business vocabulary [422]. In contrast however, common practice seems to suggest that business rules are defined separately in a very loose synchronization with business process models. Moreover, practice shows that business rules are usually defined in the form of unstructured natural language and augment other models, including business process models, in a form of comments [456]. It is obvious, however, that such unstructured business knowledge would not be useful to the latter stages of the BPM life cycle (e.g., for *execution* of the process model).

As a way to address these shortcomings, during the last decade, members of the *Business Rules community*[15] together with other scientists and practitioners tried to formalize the way business rules were to be gathered and specified [456, 458].

---

[15]http://www.brcommunity.com

However, all these efforts neither brought a significant impact on the common practice of BR specification nor did they bridge the gap between business process modeling and business rule specification [422]. At the same time, OMG joined forces with many from BPM and BR communities and in 2008 released *Semantics of Business Vocabulary and Business Rules (SBVR)* standard [359]. Although while this helped address part of the problem, that of having a way to express business knowledge in a controlled natural language via the proposed *business process vocabulary*, it did not directly address the formal integration between business rule vocabulary and process modeling diagrams. Albeit recently there has been some effort to address even these.

For instance, *Vanthienen et al.* [455] proposed an approach to implement SBVR business rules into business processes management life cycle using an SOA approach. Their architecture consisted of three layers: business rule and business process layer, service and component layer, and application layer. *Ali et al.* [25] described business rules as a separate model which is used as an integral component of BP modeling. Business processes were accordingly defined and arranged and are directly depending on the business rule model. *Milanović et al.* [330] offered to integrate BPMN with R2ML. They developed a new modeling language *rBPMN (rule-based process modeling language)*. The main idea of their work was to extend existing elements of BPMN with the BR property. *Zhao et al.* analyzed semantic programming language (SPL) to facilitate the orchestration of semantic Web services (SWS) [481]. They offered a method to integrate BR and business processes using SWS. More recently, *Agrawal et al.* proposed *Semantics of Business Process Vocabulary and Process Rules (SBPVR)* [20]. Nonetheless, these works are yet to be well adopted in the mainstream, and let alone for some, they still require undergoing thorough testing and evaluation.

Event-Driven Business Process Management (EDBPM)

In a similar line of work, the recently coined term *Event-Driven Business Process Management (EDBPM)* has also been proposed as a way of merging BPM with CEP, via "event-driven" rules. The motivation has been to enable BPM to correspond with CEP platforms via events which are produced by the BPM workflow engine and by any distributed IT services that may be associated with the business process steps. In addition, events coming from different event sources in different formats can trigger a business process or influence the execution of the process or a service, which could in turn result in another event. Moreover, the correlation of these events in a particular context can be treated as a complex, business-level event, relevant for the execution of other business processes or services. A business process, arbitrarily fine or coarse-grained, can thus be seen as a service again and can be choreographed with other business processes or services, even cross enterprises.

In many ways, the essential work, goals, and challenges in EDBPM could in fact be considered as a subset of the previously described, more general work on combining "business rules" with BPM. Albeit in the former case, the goal

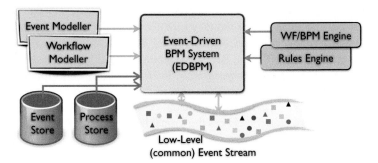

Fig. 2.6 Event-Driven Business Process Management high-level architecture

was mainly targeted toward *integrity* and *derivation* rules. For example, SBVR models rule based on the perception of elementary "concepts" (representing an entity in the business model), "fact types" (representing the relationship between the concepts), and finally "rules" which are elements of guidance that are built upon fact types [20]. For these reasons as well, the concept of the proposed "business process vocabulary" was also important. EDBPM on the other hand, as mentioned, was aimed at specifically targeting event-driven (i.e., *reactive*) rules; and as a result it thus gave rise to several event-processing languages (EPL). However, the same problems (as it did for modeling general business rules) exists in many EDBPM systems today: there is no accepted standard for such event-processing languages resulting in many different types of commercial and proprietary approaches [24, 92]; and moreover, they are not directly integrated with the process modeling languages.

A typical EDBPM could thus be illustrated as shown in Fig. 2.6, which is similar to those depicted in [457], the main point being that the various components sourced from BPM and CEP, respectively, operate almost independently, such as: event modeler vs. process/workflow modeler, event store vs. process store, rule engine vs. process/workflow engine, and process instances vs. rule instances. In fact, the only thing connecting these two systems together is the event stream at the low level, although this does not really directly benefit the process modeler (or domain experts, business users, etc.), which requires an adequate synergy at the modeling and possibly executional managing layer. In an attempt to address this, the CEP community had begun discussion of the "right standard for an EPL," albeit very controversially so far.[16]

It is rather apparent that there will be different EPL approaches for different domains, such as algorithmic trading, fraud detection, business activity monitoring, etc. Although, at present, the CEP community is gathering use cases and is classifying them according to their corresponding domains [31]. The CEP platforms

---

[16]Cf. The first CEP symposium in Hawthorne, New York, in March 2006.

come with an SQL-like language (e.g., *Coral8*,[17] *Esper*,[18] *StreamBase*[19]) or provide a rule-based EPL approach (e.g., *AMiT* by *IBM* [16], *Reaction RuleML* by *RuleML* [372], or *DRL* by *JBoss Drools Expert* [238] *and Fusion* [239]). Or some have an abstract user interface which hides any language and generates code like Java (e.g., *Tibco*,[20] *AptSoft* [200]). The models of business processes and event scenarios are deployed into a middleware platform, such as an application server, which is responsible for aspects such as high availability, scalability, transparency of heterogeneous infrastructures, etc.

JBoss jBPM[21] is one example of event-driven, Java open-source (Apache 2.0 License), and application-code embeddable BPM suites. Apart from its workflow engine that executes the process models (specified in BPEL [161] or BPMN [365]), it enables event-driven capabilities by integrating its rule engine (JBoss Drools Expert [238]), as well as its complex event processor (JBoss Drools Fusion [239]).

This effectively enables the process designer to control which parts of the process should be executed, allowing for any dynamic deviation from the process. As a result, jBPM thus claims support for adaptive and dynamic processes that require flexibility to model complex, real-life situations. However, what specifically distinguishes jBPM from other standard EDBPM suites is that it enables *stateful interaction* with the embedded rule engine. This is as opposed to simple *stateless interactions* that often only utilized the reactive rule in processes at "decision points" or data validation/enrichment at some "specific activity" node. Refer to Fig. 2.7 for an illustrative comparison.

The essential benefit of stateful rules enables utilizing much more of the rule-engine power. As a result of this, jBPM can use its DRL (rule) language to define

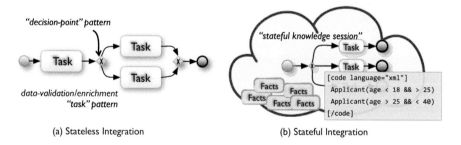

(a) Stateless Integration   (b) Stateful Integration

Fig. 2.7 Integration of rule engine with BPM systems: (**a**) stateless versus (**b**) stateful (knowledge-session) approach

---

[17]http://www.coral8.com

[18]http://esper.codehaus.org/

[19]http://www.streambase.com/

[20]http://www.tibco.com/services/standards-support/j2ee

[21]http://www.jbpm.org/

the conditions that need to be evaluated inside a gateway, thus empowering the benefit of "contextual decision" rather than just simple static data validation. For example, if we use the DRL syntax to write XOR gateway conditions, we can decide not only based on process variables, but we can also make decisions based on the *session context* that we have available. This flexibility opens a very useful set of patterns that can be leveraged by this feature. When we use the DRL language to express conditions in each of the outgoing sequence flows in the XOR gateway, we are actually filtering the *"facts"* available inside the stateful knowledge session, as shown in Fig. 2.7b. In this case, we are not filtering the information that is inside the process, we are in fact also able to check the available facts in the context to make a decision "inside" our business process. Although to enable this, an instance of the process needs to be made available inside the rule engine.

2.4.2   *Unstructured (Ad Hoc) Processes*

Unstructured processes can generally be described as complex, nonroutine business processes that are predominantly executed by an individual or a small group in a *dynamic* fashion as well as *knowledge centric*, such that it is highly dependent on the interpretation, expertise, and judgment of the humans doing the work in order to attain successful completion and that in some cases may only become available at some point "during" execution [362]. Unstructured process thus inevitably requires a great deal of *flexibility*; it also requires ample *collaboration* support, as these types of processes may heavily depend on the fluid deployment of teams exploring multiple opportunities [128].

In fact, the significance of unstructured processes are becoming increasingly important, since as mentioned earlier in this chapter, most realistic end-to-end processes are in actuality a combination of *structured* and *unstructured* processes. Moreover, research indicates that workers doing unstructured work are typically the highest paid in an organization, as these types of processes tend to be more heavily focused on the work that defines an organization and ultimately directly contributes to the companies' competitive prospects [49, 246]. Therefore, given that the majority of business process improvement investments have been focusing on structured work, this means much of the money spent on traditional process improvement does not help to most effectively reduce costs or make employees more productive [362]. Unstructured work activities thus pose many hidden opportunities for cost savings or cost avoidance.

Compared to what traditional BPM systems have provided for structured process support, including modeling, management support, as well as making enabling monitoring and measurability over processes, unstructured processes on the other hand have hardly received much focus in process improvement efforts to date, and certainly not in relation to how pervasive they are in the day-to-day activities. Rather, the primary support offered to date, for unstructured processes, is usually

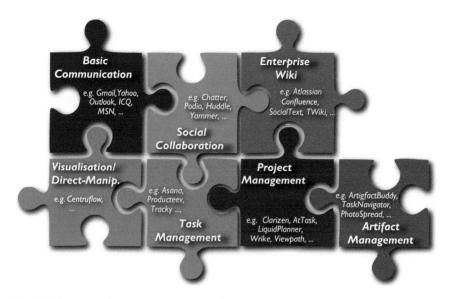

Fig. 2.8 Unstructured process-support landscape

by way of Web-based SaaS (Software-as-a-Service) tools, especially those oriented with Web 2.0 community functionality to support the required level of collaboration.

In Fig. 2.8, we have identified a range of *elementary* categories by which such unstructured process-support tools could be classified. We refer to these categories as "elementary," since while on some occasions tools may strictly fit into only one category, often enough tools adopt a variety of different concepts which are combined in order to produce a solution. In the following we have surveyed a variety of such tools from different categories. Nonetheless, while many of these tools may prove useful, they can only be aimed at selected fragments or portions of a process, whereas, from a more end-to-end perspective, since an overall typical process, as mentioned, consists of the spectrum of structured to unstructured activities, it still leaves a lot of the work to happen "outside the process" which thereby gives rise to "shadow processes" that need to use technologies such as email to manage the various process fragments, such as to exchange information and share decisions.

Basic Communication Tools

As a starting point, it is worth mentioning some of the foundational tools and services used for Web-based communication. In fact, this could refer to two main categories: *email* systems, such as: *Lotus Notes*, *Gmail*, *Outlook*, etc., as well as *instant-messaging* systems such as: *IRC*, *ICQ*, *AOL*, etc.

Social-Collaboration Tools

The proliferation of social-networking software and services has significantly transcended the basic static communication capabilities. It has enhanced it to enable a *real-time* "collaboration" space by allowing to build relationships among people with the opportunity of interacting and sharing common interests and goals. In turn, social networking has given rise to "socially enabled" software. For example, in its application to "process enactment," the concept of *"social BPM"* [104, 363] (to mean "socially enabled processes") is becoming increasingly relevant and popular. Socially enabled processes mimic the way that work is performed from an end-user perspective and experienced from a user party perspective to harness the power of continuous collaboration [91].

In fact more generally, *Johannesson et al.* highlight the particular comparison between "nonsocially" and "socially" enabled environments, being: "external authority vs. voluntary participation," "definite endpoint vs. open-endedness," "major efforts vs. quick contribution," and "access control vs. transparency" [87]. This means that a distinct principle of most socially enabled software tools is their design as a platform "encourages" but does not "enforce" the way work is done via social interaction [368]. As *Wenger et al.* phrases this, "sociality cannot be designed; it can only be designed for" [90]. Although classical IT applications are designed for a special purpose and with a special use or business process in mind, sociality acknowledges that only the users and process owners in their special (ad hoc) situations can drive the specifics of the process [87]. Therefore, instead of enforcing some ideal and predetermined process, socially enabled platforms support users with basic functionalities and a high degree of freedom [245]. Owing to these characteristics, several authors have in fact argued that ad hoc and unstructured processes show many parallels to social-software characteristics and the reason why they are commonly suggested as tools to support unstructured processes [73, 245].

Salesforce Chatter[22] is an example of a Facebook-style social-networking platform with the privacy and security essentials for an *enterprise* collaboration platform. It combines effectively with the other elements of the *Salesforce* platform that provides dashboards and other cloud-based analytics and reporting apps in order to make business intelligence (BI) more dynamic. Chatter thereby provides real-time collaboration capabilities that lets users instantly share valuable information and track important data via real-time news feeds or updates. Its automated notification system proactively alerts users about activities or materials that are relevant to their work. More importantly, from a management perspective, data distribution can be controlled via Chatter's admin capabilities to ensure data security and privacy. For example, Chatter illustrates how any authorized user can create new analysis on demand and immediately share their new charts via Chatter with peers, managers, or teams.

---

[22]https://www.salesforce.com/chatter/overview/

Enterprise Wiki Tools

As the next step toward embracing the Web 2.0 paradigm, albeit for personal, enterprise-wide, and customized use, the *enterprise wiki* acted as a publishing site for sharing and updating large volumes of information across an enterprise—as much as the same as it did for information across the Internet (cf. Wikipedia). If an organization needed a large, centralized knowledge repository that is designed to both store and share information on an enterprise-wide scale, this type of solution would be appropriate. Moreover, the methodology of this solution encouraged many-to-many communication and contribution, as opposed to the traditional publisher-consumer model. *TWiki,*[23] is an example of simple open-source enterprise wikis in addition to serving as a Web-application platform. However, most enterprise wikis provide advanced features such as: analytics, scheduling, and graph/data visualizations. Recently, many enterprise wikis have gone even beyond this, combining both advanced wiki features with social-networking features.

Atlassian Confluence[24] and *SocialText*[25] are examples of advanced enterprise wiki platforms that combine social-networking features, including real-time communication, people and group management, as well as microblogging. However, they go beyond simple wiki features and provide widgets and apps targeted toward specific business and activity needs and therefore allow much richer content creation. Confluence refers to this as the Atlassian Marketplace,[26] which provides hundreds of add-ons, including custom themes, diagramming tools, and workflow management solutions. Moreover, it allows users to build their own and share with others.

Task Management Tools

The recent *Web-oriented* and *socially enabled* task management tools are proving increasingly useful for small- to medium-size project activities [251, 256]. Essentially, the scope of these types of software has been to bring the traditional "task list" or "to-do list" concept to the cloud, which includes monitoring and tracking the status of tasks, setting and managing milestones, organizing/structuring tasks, creating reports, etc. Moreover, "socially" oriented task management applications further offer the opportunity to centralize tasks for the entire organization and enable people management over the tasks while providing a familiar format for the diverse audience of users. Authorized users are then freely able to concurrently interact with the workspaces, projects, and task lists and also collaborate about tasks with each other.

---

[23]http://twiki.org/

[24]https://www.atlassian.com/software/confluence

[25]http://www.socialtext.com/

[26]https://marketplace.atlassian.com/

Asana,[27] *Producteev*[28], and *Tracky*[29] are examples of popular task management tools, albeit there are many others. Common features among these tools include: secure workspaces that can be locked down to teams or departments, projects (lists of task items) that reside in these secure workspaces, scheduling tools (date, time, and reminder) for tasks, audit trail for tasks (date and time added and then modified by the assignee), tagging of tasks, sharing of tasks and task lists, favorite particular task(s) in a project, comments and feedback on tasks, email reminders about tasks, and integration with other popular Web tools, such as *Google Calendar.*[30]

Project Management Tools

While sharing many similarities with *task management* (as previously discussed), "project" management usually involves a more advanced set of features targeted at managing the "global" aspects of the "overall" project, rather than just individual, normally short-lived tasks [416]. For activities that require achieving large-scale initiatives, or even many smaller concurrent projects, simple task lists with due dates won't suffice. Rather, a project manager assigned to overseeing the global project (or portfolio of projects) would generally require insight into aspects such as: task estimation (length of each task, length of the overall project), project scheduling (assigning and tracking work items), resource and capacity planning (managing, monitoring, and maintaining resources to complete the project specs), portfolio-wide management (recognizing and responding to priorities), and risk management (assessing progress and identifying if attention is required) [328, 416].

Clarizen[31] and *AtTask*[32] are examples of project management software packages that resemble some of the traditional suites, such as *Microsoft Project [316]*, particularly in the way of scheduling and the way users are asked to define start and end dates for each tasks as well as set up a series of predecessors or dependencies in order to "build" the required schedules.

LiquidPlanner[33] is an example that provides an alternative in contrast to the previous traditional approaches. It enables automatically generating schedules based on the priority order of specified tasks and based on whom they are assigned to.

---

[27]https://asana.com/

[28]https://www.producteev.com/

[29]https://tracky.com/

[30]https://www.google.com/calendar/

[31]http://www.clarizen.com/

[32]http://www.attask.com/

[33]http://www.liquidplanner.com/

Wrike,[34] *Viewpath*[35], and **Redbooth**[36] are yet other examples which focus on collaboration features such as email and social-networking integration and include real-time stream management and aggregation features for a more productive overview of the project delivery.

Document/Artifact Management Tools

Document management concepts play an important role in the context of process enactments. It is clear that many processes (both structured and unstructured) produce or consume process-related "artifacts" [332]. However, particularly in the case of unstructured and ad hoc environments, where the processes are largely human driven, it becomes even more important to adequately support artifact management as first-class citizens. Traditionally, document or artifact management could be thought of as structured data repositories, usually also with versioning support. We have seen an emergence of such tools both in the Desktop domain, such as *Google Desktop Search* [186] and *MSN Desktop Search*[37], which provide a way to better manage local artifacts, and Web-based and cloud-oriented options such as *Google Drive*[38] or *Zoho Docs*,[39] which, more so, incorporate real-time and concurrent collaborative authoring.

However, when considering the intersection between document management and process enactment, there are in fact several solutions proposed for integrating knowledge management with process support, such as *EULE* [397, 398], *OntoBroker* [424], *WorkBrain* [462], *PreBIS* [135], or *DECOR* [14]. However, these tools primarily focus on *static* processes (in contrast with *weakly* structured processes), with regard to proactive information delivery; hence, they rely on structured task representation and ontologies [229]. Rather, in the case of ad hoc and unstructured processes, we have identified the following *two* main criteria for advanced support:

(a) **Artifact discovery and awareness** relate to the visibility and access to process-related artifacts by process works, during the process. Such artifacts may include documents and graphics that process workers relate with during the course of their work. Effectively, these artifacts serve as boundary objects around which communication, collaboration, and shared work take place. For example, *Whittaker et. al.* [471] found that over half of all casual interactions in an office involved some form of document sharing, where documents were mostly used as a cue or conversational prop. As detailed in the work

---

[34]https://www.wrike.com/

[35]https://www.viewpath.com/

[36]https://redbooth.com/

[37]http://toolbar.msn.com/

[38]https://developers.google.com/drive/

[39]http://www.zoho.com/docs/

by Robert Kraut et al. [270] and Steve Whittaker et al. [471] and summarized in [438], being aware of such artifacts is valuable for many reasons, including: monitoring progress, coordinating joint activities, triggering interest by seeing another person's activity, determining availability of resources, etc.

ArtifactBuddy [194] is a system that supports artifact awareness, comparable to the way interpersonal awareness is supported using an unaltered, commercial instance-messaging (IM) system. This means, just as interpersonal awareness involves knowledge about one person's up-to-the moment status, artifact aware- ness treats documents as a first-class instance-messaging "buddy"—giving each shared artifact an interactive presence, including: providing awareness cues about its editing status and notifications when new versions are committed. The system maintains a simple subscription model, between users and artifacts. In this manner it knows which people are interested in status updates of some artifact and notifies these individuals about its state. Behind-the-scenes monitoring is implemented using Windows Hooks API to monitor keyboard activity to detect if the file is being updated. Moreover, a user can converse directly with an artifact, using a set of predefined commands, in order to "pull" status updates—rather than waiting to be notified. This tool is relevant during ad hoc style work, as it combines with group communication and collaboration over the IM system; however, as typical ad hoc work may also be performed outside the system (e.g., email, human task platforms, etc.), artifacts and other activities in this case would not be captured. The user is thus confined to the limitations of the system.

(b) **Artifact contextualization and reasoning** relate to the ability for tools to be proactive in automatically providing the right information in order to empower a work context. The benefits of this approach may manifest itself in the following ways: For instance, a recognizable context could assist in providing the right information to the right person at the right time, unobtrusively sharing only what is relevant, thus guiding the process workers as they perform their tasks and drive the process. Another example might be empowering better reasoning about data, since assigning a context to data (e.g., metadata or tags) could enhance expressivity of functions and automation capabilities.

TaskNavigator [229] is an example in this category that integrates a standard task-list application with a document classification system. This results in allowing a task-oriented view over an office workers' personal knowledge spaces in order to realize a proactive and context-sensitive information support during daily, knowledge-intensive tasks. In particular document and artifacts can be structured and curated as part of the document repository, which enables users to elaborate the information model by creating new or rearranging existing structures, building relations between concepts, and assigning documents to several concepts. These structures can then be used for a contextual search (as well as in combination with keyword search). Moreover, this system realizes a proactive delivery approach, such that it automatically retrieves potentially relevant documents from various different information sources and suggests these documents to the user—this thus proposes to enhance the process participants' productivity, as they can easily identify (or automatically have these suggested to them) only the document artifacts that are relevant for a given task/activity context.

PhotoSpread [249] is another example in this category, albeit specialized for the sake of organizing and providing powerful analytical and reasoning capabilities over "photos." It does so in an interesting manner by adopting and extending the spreadsheet paradigm, whereby in addition to atomic data values, like strings and integers, "objects" (i.e., *photos* or *groups of photos*) can be displayed within any cell. Furthermore, photos can be annotated in accordance with a set of metadata, plus the ability to "tag" photos. Users can then still use conventional spreadsheet-like formulae, as well as existing functions, in order to present advanced customized views over photos, thus providing a relatively simplified means for both contextualizing and reasoning about this type of data. However, even more so, this tool enables methods for semiautomatically tagging photos: this is done using the drag-and-drop feature which updates/changes the tags, thus allowing to reorganize the photo, perhaps in a case where it was not correctly tagged. Similar methods may also be used to allow users to edit existing photo metadata as well. Effectively, this feature allows users to assign *multiple* attribute-value pairs to groups of photos with a *single* action (in contrast to more standard tagging-interfaces such as *Flickr*, which is usually limited to tagging only one photo at a time, or a group, albeit with only one tag at a time).

Visualisation/Direct-Manipulation Tools

Information management and the flow information are critical factors in processes; in particular it is assumed that if these factors are handled conveniently, it promotes dramatic increases to productivity [188]. An easily perceived method to support this is via "visualization" techniques. For example, when plain "data" is visualization, it is often thought of as transformed to "information" and "knowledge." Visualization provides an interface between the human mind and the computer, as humans are obviously much better to perceive and utilize large and complex datasets when employing visual techniques. As a result there have been several Web-based visualization toolkits that often play an important role during the ad hoc process enactment, especially when required to interact with human participants of the process. Such frameworks include: *Google Visualization API*,[40] *Cytoscape*,[41]

---

[40]https://developers.google.com/chart/interactive/docs/reference

[41]http://www.cytoscape.org/

Protovis,[42] *TufteGraph*,[43] etc. Others, including *InfoVis*,[44] *Gephi*[45], and *Walrus*[46], also provide *interactive* capabilities. The benefit of enabling interactivity (in addition to visualization) allows for constant human feedback. This could therefore be used as direct-manipulation techniques to drive customized visualization, especially when dealing with large amounts of data where the user needs to give directions for targeting the specific data they are interested in. In more advanced cases, interactivity techniques could also be used to drive the processing (e.g., visual operations could be represented as events and linked with event-processing systems).

Centruflow[47] is a software tool which is designed to offer a way for distributed users to specify relationships between data from multiple data sources using a dynamic and highly interactive "visual" techniques. At its core, the tool utilizes mind map visualization concepts, as this is particularly effective for the visualization of large amounts of structured data. However, this tool adds an addition of a direct-manipulation layer, allowing users to simply draw box/circles around information nodes in the mind map, thus creating customized clustering, as well as appending semantics using comments and tagging. The aim is for incremental structuring of data in order to make it more understandable for collective users of the data.

2.4.3 Case-Management (Semi-structured) Processes

Semi-structured processes refer to a third category of process types and exhibit characteristics from both structured and unstructured process paradigms. For instance, while they may require a certain level of structure, they nonetheless require a degree of flexibility and support for unpredictability, comparable to that of an unstructured process. Although there is yet to be accepted any formal definition, they require a subcategory of their own due to the practical significance they impart. In particular, while these types of processes may not be entirely repeatable, there are often *recurring elements* or *patterns* that can be reused [130, 423]. For this reason, these types of processes are therefore also commonly referred to as "case-based" processes—where we could think of a "case" as a variation to a family/pattern of processes.

Traditionally, workflow management technology has been applied and relied upon in many enterprises [166, 289, 450]. However, there appears to be a significant

---

[42]http://vis.stanford.edu/papers/protovis

[43]http://xaviershay.github.io/tufte-graph/

[44]http://philogb.github.io/jit/

[45]https://gephi.github.io/

[46]http://www.caida.org/tools/visualization/walrus/

[47]http://www.centruflow.com/

gap between the promise of workflow technology and what systems really offer [9]. As indicated by many authors, workflow management systems are too restrictive and have problems dealing with change [19, 101, 158]. In essence, workflow and process management systems are "single" case driven—this means they only focus on a single process instance. Rather, the only way to handle several cases in parallel would mean to describe different business processes. However, from the viewpoint of the workflow management system, these cases are logically independent. Thus, effectively each case is handled as *one* workflow instance in isolation with others.

In contrast, what case management requires is that there is no predetermined sequence. Unlike traditional workflow and business-process systems that require the sequence and routing of activities to be specified at design time (as otherwise they will not be supported by the system), case management requires the ability to add new activities at any point during the life cycle of the case as the need for them arises [130]. However, the technology to handle this is yet to be able to support these requirements [423]. As a result, case management has often only intensely been managed manually, typically paper driven, thus highly susceptible to delay and poor visibility, particularly where isolated parts of the process are automated independently where possible, by stand-alone tools, or worse so by legacy systems.

2.4.3.1 Common Characteristics of Case Management

The participants of case-management process (also referred to as "case workers") typically need to manage a complex set of tasks from process initiation until completion. This usually requires the interaction between other participants and external services/tools plus requires the maintenance of documents and other process-related artifacts and records. In general, case-management scenarios share a lot of common characteristics [423]; we itemize below the five key ones that we have identified:

- *Knowledge intensive.* Case-managed processes are primarily referred to as semi-structured processes, since they often require the ongoing intervention of skilled and knowledgeable workers. For these reasons, it is considered that human knowledge workers are responsible to drive the process, which cannot otherwise be automated as in workflow systems. Workers usually acquire their knowledge from experience with other cases and, more so, may require to collaborate with others who are more experienced colleagues.
- *Collaborative.* As mentioned above, collaboration is an important feature; however, it may not only be relevant between colleagues, but case workers may also require coordinating collaboration (e.g., meetings/interviews) between interested parties and participants of the process. In some cases also, these different team members/participants may need to access case information and discuss it with each other. Furthermore, a related aspect is thus the adequate maintenance of participants' roles as well as the creation and adherence to the allowed rights for each type of role.

- *Diversity of Information.* The information associated with case process are usually complex, entailing the collection and presentation of many different types of documents and records. This may also be due to the diversity of participants interacting (as mentioned above). Moreover, it is thus important to make the required information easily accessible for smooth enactment of the process, albeit it may prove difficult especially in typically dynamic and flexible environments.
- *Variability.* While a particular case may share a general structure, the actual particular instance that eventuates may be somewhat (if not significantly) different from the predicted case pattern. In some case it is also "impossible" to predetermine certain aspects of the process structure. Case management must therefore make provision for cases to change, which may often occur in an unpredictable, dynamic, and ad hoc manner as it progresses through its life cycle. In some cases, although some elements may be fixed, there can still be considerable variation in how steps are executed, based on the particular circumstances of the instance.
- *Interrelated.* It is also common that separate cases may impact other case processes. Moreover, cases may be linked explicitly, or else they may also be linked by inference, and thereby conducted with this inferred link in mind.

2.4.3.2 Review of Current Approaches

Due to the current aforementioned gap in technology, the "case-management" paradigm has been recognized as a promising approach to support semi-structured processes. It can also be especially suitable to support knowledge-intensive processes referring to those processes that are largely driven by human participants and vital knowledge workers. For these reasons, recently there have been several efforts to push this, such as *Forrester analysts Le Clair and Moore* [283], suggesting the increasing importance and revival of the concept, while the *OMG* is also currently working on an according standardization. Thus at present, there is no concrete all-encompassing tool or framework that is capable to adequately support these requirements, whereas at the moment case management is yet only considered "a general approach" rather than being a "mature tool category." Nonetheless, in the following we have identified a few of the latest attempts that may contribute toward supporting case-management process:

Van der Aalst et al. [9], for example, proposed a framework that supports the user with information about what activities can be performed based on the current status of the case. Although, this suggests that the system has to be trained and possible case statuses have to be pre-modeled. While it might be useful in some contexts that are at least semi-structured, it nonetheless proves rigid where extra flexibility is required.

Hagel et al. [203] propose shifting from a push-based approach in BPM and among IT systems in general to a pull based architecture. In this way the user-driven model is a logical consequence of rising complexity and knowledge intensity. Applying this to case management leads to the proposal of *emergent case management*.

Bohringer [87] refers to *emergent case management* as the approach for bottom-up management, as opposed to the top-down approach apparent in traditional BPM systems. This work further suggests using social software (particularly tools involving: tagging, microblogging, and activity streams) and thereby aiming toward using this content in a process-based manner. It claims to empower people to be at the center of such information systems, where the goal is to enable users to assign activities and artifacts independent of their representation to a certain case, which can be dynamically defined by users.

Vanderfeesten et al. [454] outline a hybrid approach, combining activity-based and artifact-based control principles. Although there still is a logical flow, case workers can nonetheless be authorized to skip and/or redo activities. The case (and its data) is the main driver of the process; and activities can be constrained by the availability of case-data elements. Moreover, certain case-data elements can only be entered by certain case activities, where such restrictions are modeled as explicit relationships between case-data elements and activities.

2.5   Conclusions and Future Directions

In this chapter we have identified a set of dimensions by which processes and process systems can be described. While most commonly and broadly processes are often categorized in terms of *paradigm*, as we have shown this represents yet one dimension, whereby, in order to fully understand the landscape of process capabilities in different environments and contexts, a more in-depth analysis is required. Accordingly, we established three main dimensions for discussion, namely, (a) process paradigm (i.e., the type of activities that are well supported), (b) the implementation technologies, and finally, (c) the representation model or language/s available to a user. We then first examined the various identified implementation technologies, which we emphasize may often be independent from the other facets. For this we identified three main categories: (a) workflow engines, (b) rule engines, and the traditional rather brute-force approach of (c) Program-coded-based solutions that we also discussed for the sake of completion. Subsequently, we then provided an in-depth survey covering some of the popular and relevant process frameworks and support tools, where we presented separately according to process paradigm: (a) structured, (b) unstructured/ad hoc, and (c) case-management (semi-structured) processes. Within each paradigm, we then identified the various representation models/languages available.

It is apparent that, based on our discovery and analysis, although process management provides a technological revolution in the light of advancing information processing systems, there still lies a significant challenge to be addressed. In particular, although workflow and business process management systems are well suited for highly structured and procedural processes, they do not cater for ad hoc or less-structured use cases that often require flexibility and provision for ongoing customization. As a result, we identified a large variety of Web-based support tools covering a variety of areas, including: basic communication; social-collaboration; project-, task-, and document-management; as well as visualization and direct-manipulation tools. It is evident these tools play an important role in managing these types of less-structured processes; however, they often result in "shadow" or "invisible" processes, as there is no platform providing end-to-end process support; thus, they are only managed by hand with no opportunity for management, monitoring, or control.

The goal for future process-support systems will thus be to bridge this technological gap between structured and unstructured processes. Moreover, the next generation of process-support systems would also benefit from incorporating the concepts and characteristics that case management has proposed. At present, the first formative and encouraging work in this domain has been proposed by Barukh et al. [48]. We are thereby optimistic that such work, together with the motivating ideas expressed in this chapter, will provide the foundation for continued growth into a new breed of enhanced process support.

Chapter 3
Process Matching Techniques

As business process management technology is widely used, organizations create and run hundreds or even thousands of business process models. Repositories have been proposed for managing a large collection of process models, offering a rich set of functionalities, such as storing, querying, updating, and versioning process models. One of the functionalities required to manage these repositories is by comparing two process models. This chapter presents techniques for process matching. We describe some application scenarios where process matching is needed, such as behavior-based Web service discovery, scientific workflow discovery, process similarity search, etc. Some of these applications require evaluating the similarity of process models. We overview similarity measures that can be considered for different types of process attributes. Then, we present techniques that compare process models' different perspectives: interface, business protocol, and process model. Depending on the application, the appropriate technique or a combination of techniques could be used. We provide an analysis of existing techniques and discuss some open problems.

3.1 Introduction

As business process management technology is widely used, organizations create and run hundreds or even thousands of business process models. For example, there are more than 600 process models in the SAP reference model [477] and there are over 3000 models in the collection of process models for Suncorp, an Australian bank and insurance company [460]. Thus, process models are vital assets for organizations, and it is essential for these process models to be managed effectively. Some repositories have proposed for managing large amounts of process models. The repositories could offer a rich set of functionalities, such as storing, querying, updating, and versioning process models. They also provide more sophisticated functionalities, such as advanced model-based analysis, comparing models, reusing

© Springer International Publishing Switzerland 2016   61
S.-M.-R. Beheshti et al., *Process Analytics*, DOI 10.1007/978-3-319-25037-3_3

existing models to design new process models, and so on. Interested readers may refer to a comprehensive survey of the field [477].

One of the functionalities required to manage these repositories is comparing two process models. Matching process models is a critical operation in many other application domains, such as behavior-based Web service discovery, scientific workflow discovery, process similarity search, etc. For some applications, users need an exact match (e.g., for behavior-based Web service discovery at execution time). For other applications, like process model reuse, if an exact match does not exist, an approximate match should be returned to the user. Thus, a similarity measure is needed to quantify their similarity.

Comparing two processes consists in comparing all the attributes of their descriptions, corresponding to different perspectives and assessing if they are identical, equivalent, or similar. In modern enterprises and applications, business processes are executed on the Web and exposed as Web services. As Web services, they have a WSDL interface describing public operations. The constraints among public operations are defined as business protocols. Moreover, the data manipulated (messages exchanged) by the two processes have to be compared. So, comparing two processes consists in comparing the: (a) Schema of their messages (b) Interfaces (set of offered operations) (c) Business protocols (d) Process models (the executable process models).

The goal of this chapter is to present techniques useful to compare business processes. The remainder of the chapter is organized as follows. In the following, we describe applications where matching process models and similarity measures that may be useful for assessing similarity of different process characteristics are needed (Sect. 3.1.2). Section 3.2 overviews techniques used for schema matching, mostly from a database domain. Section 3.3 presents techniques for comparing interfaces, taking into account their operations and QoS factors. Section 3.4 deals with protocol matching techniques. Section 3.5 proposes a classification for process matching approaches following several dimensions (like the abstract model used for representing processes, the exact or inexact type of match) and overviews some of the recent matching solutions in light of the classification proposed.

3.1.1 Application Domains

In this section, we present several scenarios that are required to compare two process models and to assess their similarity:

Web Service Discovery and Integration Consider a company that wants to integrate a new Web service into its existing application, in order to replace the Web service offered by its current partner. Many existing services are programmed to exchange data according to a specified process model (also called business or conversation protocol). Thus, the company will search for a service that has a conversation protocol that is compatible with its protocol in order to integrate it without errors in the existing application. If such a service does not exist, the most compatible one has to be found. If the service is not fully compatible, the company will adapt its service or will develop an adapter in order to interact with the retrieved service. In both situations, the differences between the process models have to be automatically identified. In the former case, finding the most similar service allows to minimize development cost. In the latter case, identifying automatically the differences between protocols is the first stage in the process of semiautomatically developing adapters (see [66] for adapter patterns and [447] for mediation of OWL-S processes).

Retrieving Scientific Workflows   Collaboratories (a new concept designating virtual scientific laboratories) emerged on the Web, allowing scientists to share and reuse the scientific workflows describing their experiments. As an example, myExperiment[1] is one of the current public workflow repositories containing hundreds of distinct scientific workflows, contributed by hundreds of users. While current retrieval facilities in workflow repositories are mostly limited at browsing features and keyword-based searches, some works like [182, 183] elicited requirements for workflow discovery through an experiment with scientists and workflow developers. The experiments showed: (a) the willingness of users to share and reuse workflows, (b) the limits of current retrieval facilities, and (c) the need for discovery features taking into account the process model.

Retrieving Business Processes in Repository   This application is similar to the previous one, but this time the search concerns business processes. Repositories with hundreds of process models become increasingly common: reference model repositories distributed by tool vendors (like SAP) or reference models for local governments [137]. In order to retrieve relevant models from these repositories, browsing is not sufficient. Tool support is needed allowing to find models similar to a user query. This is necessary, for instance, when adding a new model, to see if there are duplicates or overlapping process models. In case of reference model repositories, such a search allows to find reference models that are related to an existing enterprise process.

Autocompletion Mechanism for Modeling Processes   Modeling business processes is a time-consuming and error-prone task. Users can be assisted by providing an autocompletion mechanism [157] that proposes subsequent fragments. Thus, the fragment already modeled has to be compared (transparently for the user) to existing templates in the repository in order to find similar processes and to recommend subsequent fragments.

Delta Analysis consists in finding differences between two business processes. For example, suppose a company wants to evolve its business process to meet an international standard. Thus, analysts need to identify the differences between the

---

[1] www.myexperiement.org

internal business process and that of the standard and ideally to evaluate the cost induced by the reengineering effort.

Version Management consists of managing different versions of business processes evolving over time. This evolution can arise from different situations, such as changing laws, proposing a special offer, exception handling, etc. Companies need to have at their disposal operators for managing process versions, which allows ensuring business process consistency and reusability. Such operators include Match, for finding correspondences among models in situations where models are developed independently [353]; Diff, for finding differences between models [279]; and Merge, for combining two models with respect to known relationships between them [353].

3.1.2 Similarity Metrics Used When Comparing Process Models

When comparing business process models, we would consider the following six aspects:

- Name similarity: Given two names (e.g., activity names), the name similarity metric returns the degree of similarity between them by syntactically comparing the name strings. The similarity computation can be done using some approximate string similarity functions, such as edit distance, Jaccard, q-grams, and so forth [207].
- Description similarity: Process models often contain descriptions in natural language to express the intended semantics of models. The description similarity metric compares these descriptions to determine the similarity between process models.
- Concept similarity: Semantic annotations can be added to process models to enrich the semantics of models, e.g., associating a predefined semantic goal with an activity. This similarity metric measures the similarity between semantic annotations by either syntactically comparing the sets of semantic concepts or logic reasoning.
- Numerical value similarity: Some QoS properties have numerical values, such as price, time, or reputation. Numerical similarity metric computes the similarity between these QoS property values. In some cases, a standardization step is necessary before computing the similarity.
- Structural similarity: This similarity metric computes the similarity between structures of process models by comparing the topology of models. As process models can be seen as graphs, the similarity between process models can be measured using some graph matching techniques, such as graph-edit distance or similarity flooding [325].

- Behavioral similarity: This metric determines the degree of similarity between the execution semantics of process models. Given a process model, the execution semantics can be described in terms of (a) the set of traces that the model can generate, (b) a labeled transition system that consists of all the states of the model and all transitions causing to change from one state to another, and (c) casual footprints [154].

Typically, similarity functions (metrics) play an important role in measuring the similarity between a pair of process models. Over the last four decades, many different similarity functions have been designed for specific data types (e.g., string, numeric, date/time, or image) or usage purposes (e.g., typographical error checking, structural similarity estimation, or phonetic similarity detection). Here, we discuss some of them, which can be used for comparing several aspects of process models, and their characteristics, which are considerable when choosing appropriate similarity functions for the purpose.

For *string* data (e.g., element names or descriptions), if strings contain variations and errors, applying the exact string matching may produce poor results. Thus, the similarity between those strings can be computed with approximate string matching functions [207], which can be roughly categorized into two groups: *character-based* and *token-based* functions. The character-based functions (e.g., edit distance, Jaro, or q-grams) consider characters and their positions within strings to estimate the similarity. Each of them works well for particular types of strings [83, 115]. For example, edit distance function [207] is suitable for comparing strings that are short and have misspellings. On the other hand, the token-based functions (e.g., Jaccard or tf-idf cosine similarity) are appropriate in situations where the string mismatches come from rearrangement of tokens (e.g., "Verification Order" versus "Order Verification") or the length of strings is long, such as long descriptions [267].

For *numeric* data (e.g., when comparing QoS), one can choose different functions for comparing numeric values, such as relative or Hamming distance. For example, if numeric values have a fixed length, Hamming distance might be good for comparing them [437]; otherwise, relative or absolute distance can be considered. Non-numeric data types, such as *date* or *time*, may be transformed into numeric values or specific formats, such as dd/mm/yy or mm/dd/yy for date attribute values. Functions used for numeric data types can be taken into consideration in computing the similarity.

Even though there is no similarity between any two data values, further comparisons can be made because of the *semantic relationships* between them [34, 141, 263, 420]. For instance, consider two strings "Verification Order" and "Order Checking" of activity labels. Although normal string comparison functions may fail to see the similarity (e.g., scores 0.33 and 0.11 obtained from comparing those two strings using Jaccard function and edit distance function, respectively), the strings still become similar to each other, if we keep the information that "Verification" is a synonym of "Checking." For bridging the semantic gap in attribute values, we can exploit the auxiliary information, such as synonym and abbreviation dictionaries, or WordNet.

For *structural similarity measure* of process model graphs, we can use graph matching algorithms that convert process models as labeled graphs and determine the similarities between nodes of the graphs. Usually, to estimate the similarity between a pair of nodes from two model graphs, they consider their positions within the graphs. The underlying idea is that the similarity between two nodes could be the similarity between their neighbors. For example, two non-leaf nodes are structurally similar, if their children are similar. Some of such graph matching algorithms [154] include graph-edit distance, A* graph-edit distance, similarity flooding, and so on. When selecting such graph matching algorithms, it might be good to understand the characteristics of the matching algorithms. For example, when we compare process model graphs, the lack of edge labels indicates that it might not be a good choice to apply the `similarity flooding` algorithm that is one of the graph matching techniques and is known as giving good accuracy in the context of schema matching. The reason is that the algorithm heavily uses the node labels as well as the edge labels [154].

To compute the behavioral similarity between two process models, we can consider several methods as follows: (a) matching generated traces, (b) simulating transition systems, and (c) comparing casual footprints [154]. Given two process models M_1 and M_2 where each model generates a set of traces, the similarity between M_1 and M_2 can be calculated as $\frac{2 \times |S(M_1) \cap S(M_2)|}{|S(M_1) \cup S(M_2)|}$, where $S(M)$ denotes the set of traces that model M produces. Another method for measuring the execution semantics is based on a labeled transition system that consists of all states of the model and all transitions between states. The similarity between a pair of process models can be computed by examining how many states of the transition systems allow the same transitions. Lastly, we can measure the similarity between two process models by considering the casual footprints which are derived from the models and represented in a vector space (inspired from information retrieval domain). Techniques to evaluate process behavioral similarity will be presented in detail in Sect. 3.5.4.

3.2 Schema Matching

Schema matching is the task of finding correspondences (matches) between elements of two schemas. The task takes as input two schemas, each consisting of a set of elements, and produces as output the matches between these elements. Examples are "price= amount" and "address= concat(street, city, state)." Schema matching is a challenging problem since only the designers of the schemas being matched can know the exact meaning of schema elements [310]. Many works [72, 74, 141, 142, 142, 285, 295] have dealt with the schema matching problem. In this section we present individual and multi-matcher matching approaches, which are two categories of major techniques to schema matching.

3.2.1 Individual Matcher Approaches

This category of approaches compute matches between elements based on a single matching criterion. They employ machine learning [72, 136, 142, 293], clustering [309], statistics [221, 310], or rules and heuristics [71, 310, 335]. The approaches use a particular kind of schema information, such as element names, data types, or relationships, between elements. We distinguish them into element-level and structure-level techniques, based on what kind of data the techniques work on [390]. An element-level matcher finds matches between schema elements without considering their relationships between other elements as follows:

- *Linguistic-based matching.* Linguistic-based matchers consider the similarity of element names and their textual descriptions to identify semantically similar schema elements. Similarity between element names can be computed using similarity functions, such as edit distance, n-grams, or prefix/suffix [141, 309, 325]. Semantic relationships (e.g., synonyms, abbreviations, or hyponyms) between element names could be considered to match elements, using common knowledge or domain-specific thesauri.
- *Constraint-based matching.* Constraint-based techniques consider the internal constraints, which are specified when defining elements, such as data types, cardinalities, value range, or foreign keys. These constraints can be considered to determine the similarity of schema elements. For example, element `BirthDay` of `date` type could be matched with element `Born` of `date` type by considering the data type, even if the element names are not similar to each other.

A structure-level matcher determines matching elements by analyzing how they are similarly structured, such as having similar relationships with other elements. The entity examples, which have graph-like structures containing elements/terms and their relationships, include: processes, DB/XML schemas, taxonomies or ontologies. Some approaches [141, 309, 311, 325] rely on graph matching algorithms that convert input schemas as labeled graphs and measure the similarities between nodes of the graphs. For example, in [311], the authors compute the similarity between nodes using their relationships with other nodes.

3.2.2 Multi-matcher Matching Approaches

While individual matchers rely on a single matching criterion to find matches, multi-matchers [74, 136, 141–143, 159] combine several base matchers where each base matcher considers only one aspect of schema elements. For example, a name matcher computes the name similarity by using some similarity functions (e.g., edit distance or Jaccard), while a type matcher compares the data types of elements using a data type compatibility table. Multi-matchers can be distinguished into *hybrid* and *composite* matchers [309]. Hybrid matchers [70, 292, 335] leverage different criteria to find matches, while composite matchers [141, 142] independently execute several matchers, including hybrid matchers, and combine their results to make a final matching decision. Some examples of hybrid matchers are:

- *Transcm*: The Transcm system [335] relies on schema matching to automatically translate a source schema instance to a target schema instance. The system exploits names and structures of schema elements to perform matching.
- *SEMINT*: The SEMINT system [292, 293] exploits neural networks to determine matches between elements. It can select multiple match criteria, up to 15 constraint-based and 5 content-based matching criteria [390].
- *Cupid*: Cupid is a hybrid matching technique that integrates linguistic and structural schema matching algorithms. The technique is performed in three phases. In the first phase, it measures the linguistic similarity between element names. The second phase computes the structural similarity of schema elements by considering the locations where the elements occur in the schemas. The third phase computes the weighted sum of both linguistic and structural similarity scores to decide final matchings.

Some examples of composite matchers are:

- *LSD*: The LSD system [142] uses machine-learning techniques to perform matching. In addition to a name matcher, a number of base learners (matchers) trained using data sources are used to match a new data source to a mediated schema.
- *COMA*: COMA [141] is a composite matcher system that provides a library consisting of individual matchers, hybrid matchers, and one reuse-oriented matcher. The system allows to combine multiple matchers in a flexible way.

Some approaches [285, 295] proposed techniques to automatically tune composite matcher systems. Given a matching task, they select the right matchers to be combined and adjust various parameters (e.g., thresholds, weights, coefficients) of the matchers, which might have a large impact on the match results. In [285], the authors have developed eTuner, a system for self-tuning schema matching systems. Given a schema, the system first generates synthetic schemas including the given schema, for which the ground truth matching is already known. Then it uses the synthetic schemas to find the best tuning of matchers. In [295], the authors proposed an approach (called RiMON) for automatically determining parameters of matchers for ontology matching. They considered two similarity factors: textual similarity and structural similarity factors to combine matchers.

3.3 Interface Matching

Service specifications refer to the descriptions of syntactic or semantic aspects of a service. They are publicly available to potential clients for (a) enabling developers to implement the client services that can correctly interact with a service and

(b) allowing the selection of services that satisfy clients' requirements, either at development time or at runtime. In today's Web, service specifications typically include service interface descriptions and business protocol descriptions [384]. A service interface (syntactically specified in WSDL) defines which operations are provided by a service, along with message formats and data types, while a business protocol specifies which message exchange sequences are supported by a service [68].

3.3.1 Matching Operations

Matching service interfaces can be achieved by considering multiple and complementary sources of evidence like: their labels, inputs, outputs, textual descriptions, and operational properties (like execution time, cost, etc.). Existing approaches can be broadly divided into four categories: (a) syntactic- and linguistic-based approaches, (b) ontology-based approaches, (c) signature matching-based approaches, and (d) adaptation-based approaches.

In the syntactic -and linguistic-based approaches, the similarity of service interfaces is assessed using the label of their operations. In closed environments where the services are developed on the same tool using a shared vocabulary for naming their services (e.g., RosettaNet PIPs[2]), the similarity of the services is quite simple since it is based on the equality of the service name identifiers. However, this task can be quite complex if they are designed by different modelers, using different vocabularies, as is often the case in the open environment of the Web and in the service-oriented architecture. In real life, service names are often formed by a combination of several words and can contain abbreviations. Thus, the similarity of their labels can be measured using different string matching techniques (string-edit distance [287], *n-gram*, *Check synonym*, *Check abbreviation*, tokenization, and stemming) that can be combined to design sophisticated similarity measures [138, 374]. The *n-gram* technique estimates the similarity between two strings according to the number of common *qgram* between activity labels [32]. The *Check synonym* technique uses a linguistic dictionary (e.g., Wordnet [333]) to find synonyms, while *Check abbreviation* [197] uses a custom abbreviation dictionary.

Within the framework of semantic Web, service input/output attributes can be annotated [374] using some description logic-based ontology languages (e.g., OWL-DL [230] or F-logic). In such settings, the approaches retrieve the services that semantically match with a given service request, based on description logic reasoning methods. They use the terminological concept subsumption relations computed in the underlying ontology [163, 435]. Some examples of such logic-based approaches are *OWL-S-UDDI* matchmaker [434], MAMA [120], *RACER* [291], and *WSMO* [282]. For example, in [65, 434], the similarity of two services is

---

[2]http://www.rosettanet.org/

assessed with regard to the semantical relationships occurring between their inputs and outputs, e.g., checking whether they have the same type or one is a specialization of the other. In [255], several filters are defined for service retrieval: the name space, the domain of ontology that is used, and types of inputs/outputs and constraints. The *OWL-S-MX* matchmaker [435] extends this approach by introducing approximate matching metrics based on information retrieval similarity metrics applied to features characterizing concepts in the ontology. They show that the hybrid matching techniques significantly outperform logic-based-only approaches, under certain constraints.

Some work proposed techniques that allow combining multiple similarity measures in order to develop tunable approaches [97, 146, 436]. For instance, the approach presented in [436] combines domain-dependent ontological information with domain-independent semantic for improving service matching quality. The domain-independent relationships for common terms are derived using an English thesaurus, while specific relationships between industry and application-specific terms are inferred using a domain ontology.

The signature matching-based approaches focus on finding similar operations to a given function or service operation signature from a repository of software components [479, 480] or Web services [146]. In [479], to find software components matching with a given component, the authors propose an approach for comparing the signature of software components by considering the parameter name, parameter type, parameter order, and so on. In [480], the authors formally define the problem of matching software components by considering signature matching and program behavior matching. Their approach is based on the analysis of data types and post-conditions.

An approach for discovering services similar to a given query or operation in a collection of services is proposed in [146]. The approaches in service matching can rely on more information than that in component matching as service interfaces are more expressive than software signatures. Instead of computing the exact matching between schema elements of messages, the approaches measure the similarity between them, based on IR techniques. In addition, their underlying algorithms are based on the learning and statistical analysis methods (e.g., clustering) of service interface descriptions. For example, in [146], the authors propose an approach for supporting the operation similarity search that finds service operation similar to a given one, by employing a clustering algorithm grouping parameter names into meaningful concepts.

The adaptation-based approaches [172, 385] focus on adapting WSDL interfaces to incompatible clients. Matthew Fuchs [172] presents an approach for generating service views on top of WSDL interfaces by altering WSDL interfaces. Thus, interactions with incompatible services are enabled with the views. To generate such views, they use a scripting language that specifies the changes of WSDL interfaces. The commands of this language are used for inserting, deleting, and updating some components of WSDL interfaces. In [385], the authors assume that a common base interface can be used for deriving the interface of all services that provide a similar functionality. The derivation can be done by using a set of primitives that allow to add parameters to or remove parameters from operations. They also propose techniques for addressing four types of incompatibilities (i.e., structural, value, encoding, and semantic) that arise during interacting with other services.

3.3.2 Matching QoS Factors

Another stream of research integrates the operational properties, which are name designated as QoS (quality of service) properties, in the matching process. QoS for Web services is a broad concept that encompasses various nonfunctional aspects of services, such as performance, availability, reliability, reputation, and so on. QoS matchmaking is a process that takes as input an inquiry's QoS requirement and produces as output a set of published advertisements that meet the inquiry requirement. The QoS matchmaking requires both the description that specifies services' QoS properties and the corresponding matchmaking algorithm.

Several approaches have been proposed for describing the QoS properties of services [273, 308, 320, 360, 392, 443]. In [392], the authors present a new model for extending UDDI to describe the QoS properties about a particular service. In their model, service providers need to describe the functional and QoS aspects of their proposed services. The extension is made on top of the UDDI model so that their approach only supports the syntactic matchmaking of Qos terms and specifications. In [320], the authors propose a conceptual model for describing Web service reputation. Their conceptual model consists of a general component, such as attribute types and attributes (e.g., price, reliability), and domain-specific components, which apply to a specific service. Their model does not represent some other concepts, like QoS offers and demands. In [443], the authors propose a language, Web Service Offerings Language (WSOL), for formally specifying various constraints and classes of service for Web services. Their language is an XML-based notation compatible with WSDL.

A service offering described in WSOL contains various properties, such as functional properties (e.g., preconditions and post-conditions) and QoS properties (e.g., performance, reliability, and availability). They argue that different classes of services should be offered to satisfy a large number of and different kinds of customers and to successfully handle the situations where there are some variations in QoS because of network problems and other reasons. In [360], the authors propose an approach for defining and matching WS agreements. Their approach supports the description of QoS properties and their matching using user-defined rules. In [308], the authors propose a model for modeling QoS features related to the dynamic nature of service environments, such as end-user mobility, adaptability, and context awareness of application services. The model addresses QoS on end-to-end basis by considering QoS features of all the resources and actors involved in fulfilling services, e.g., network and hardware resources, application services, and end users. For further details of the QoS description methods, see [273] for a recent survey.

Some approaches [271, 272, 428, 453, 482] have been developed for matchmaking QoS specifications. In [482], the authors propose an approach for extending the DAML-S Web service description language to include a QoS specification ontology. The QoS ontology consists of three layers: the (a) QoS profile layer defined for QoS matchmaking purpose, (b) QoS property definition layer for defining QoS properties and their domain and range constraints, and (c) QoS metric layer for defining QoS metrics and their measurement. They also present a QoS matchmaking algorithm based on the concept of QoS profile compatibility. The matchmaking algorithm relies on a description logic reasoner that computes the subsumption relationship between a request and all available QoS advertisements. In [317], the authors exploit a symmetric QoS model that represents mathematical constraints for QoS metrics and user preferences. Before matchmaking, they transform a QoS specification into a constraint satisfaction problem (CSP) [453] for checking consistency, i.e., to compute whether the corresponding CSP, which consists of all the mathematical constraints, is satisfied or not. Then, they perform matchmaking based on the concept of conformance, i.e., to check whether every solution for the CSP corresponding to an offer can be also a solution for the CSP corresponding to a demand. After matchmaking, two results are generated: one for matched offers with respect to the demand and the other for not matched offers. In [428], the authors propose an approach based on an attribute constraint definition. Their approach generates interval records by computing the combinations of the possible intervals for all the attributes. Here, an interval for a particular attribute defines the lower and upper bound of the values of the attribute. Their matchmaking algorithm uses these interval records to determine whether two constraint descriptions are compatible.

In [271], the authors propose an ontology-based approach for QoS-based Web service description. First, they present an upper ontology for describing QoS of Web services (called OWL-Q) that specifies the possible parts of QoS metrics and constraints. They design the ontology into several facets that can be easily extended. In addition to the upper ontology, they present a mid-level ontology used for defining all domain-independent QoS metrics. This mid-level ontology is a basis for defining new QoS metrics on low-level ontologies. They also develop a semantic QoS metric matchmaking algorithm that computes the similarity between two QoS metrics. They incorporate their semantic matchmaking algorithm with syntactic QoS-based matchmaking algorithms to obtain better results in terms of precision and recall. In [272], the authors present a set of requirements that should be met by QoS-based description models. They introduce and analyze which requirements are required for performing an effective and user-assisting QoS-based matchmaking process. They also provide a proposal on the development of a semantically enhanced QoS broker by extending the current functional Web service registries.

3.4 Protocol Matching

The rapid growth of Web services has led to the proliferation of functionality-wise equivalent services, but having differences in their descriptions, and therefore has given rise to the need for service adaptation. For services having multiple and dependent operations, besides the interface, their business protocols have to be compared also. Works on adapter development (e.g., [66] by the authors and also some later work by Dumas et al. [153]) have identified mismatches commonly occurring between Web service protocols.

To describe common mismatches at the protocol level, we use a supply chain example. Assume that protocol P_r of service S_r expects messages to be exchanged in the following order: clients invoke login and then getCatalogue to receive the catalog of products including shipping options and preferences (e.g., delivery dates), followed by submitOrder, sendShippingPreferences, issueInvoice, and makePayment operations. In contrast, protocol P of the client allows the following sequence of operations: login, getCatalogue, submitOrder, issueInvoice, makePayment, and sendShippingPreferences. As service S_r does not charge differently even if the shipping preferences differ, clients can specify their shipping preferences at the final step. Based on [66, 153], we classify protocol-level mismatches as follows:

- *Ordering Mismatch*: This mismatch occurs when protocols P and P_r support the same set of messages but in different orders, as shown in the above example.
- *Extra Message Mismatch*: This mismatch occurs when one or more messages specified in P_r do not have any correspondence in P. In the above supply chain example, assume that protocol P_r sends an acknowledgment after receiving message issueInvoiceIn, but protocol P does not produce it.
- *Missing Message Mismatch*: This mismatch is the opposite case of the Extra Message Mismatch, i.e., one or more messages specified in P do not have any correspondence in P_r.
- *One-to-Many Message Mismatch*: This mismatch occurs when protocol P specifies a single message m to achieve a functionality, while protocol P_r requires a sequence of messages $m_1, \ldots, m_n$ for achieving the same function. Suppose protocol P expects purchase order together with shipping preferences in one message called submitOrderIn, while protocol P_r needs two separate messages for the same purpose, namely, sendShippingPreferencesIn and submitOrderIn.
- *Many-to-One Message Mismatch*: This mismatch occurs when protocol P specifies a sequence of messages $m_1, \ldots, m_n$ to achieve a functionality, while protocol P_r requires only a single message m (which can be created by combining messages $m_1, \ldots, m_n$) for the same function. In this type of mismatch, an assumption is required in which all the messages to be aggregated need to be issued consecutively, i.e., there is no other actions waiting for other messages to be sent during this message aggregation. This assumption is required in order to avoid deadlock.
- *Stream-to-Single Message Mismatch*: This mismatch occurs when protocol P_r issues a stream of messages until a condition is satisfied, while the protocol

P only requires one message to be sent. In the example, assumed protocol P_r sends shipment notification incrementally until the products are delivered, while protocol P only sends a single shipment notification.

• *Single-to-Stream Message Mismatch*: This mismatch is the opposite case of the Stream-to-Single Message Mismatch, in which protocol P_r issues a single message, while protocol P sends a stream of messages until a condition is satisfied.

The authors of [356] propose protocol-level analysis through simulating the interactions between two protocols P_s and P_c. In this process, they explore all possible message exchanges between the two services according to P_s and P_c and identify the mismatches between them. This is akin to the adapter generation process, and that is why we refer to it as adapter simulation process. The purpose of this process is threefold: (a) identifying plausibility of interface matching results by considering the full interactions between protocols, (b) generating adapter rules for mismatches that do not result in deadlock, and (c) identification and analyzing mismatches that result in deadlock to examine if it is possible to provide the required interface mappings, and rules, that resolve the deadlocks.

Protocol information can be used also during the interface matching of two services for adapter development. The approach presented in [357] proposes to use protocol information and type of adaptation (for replaceability or compatibility) in order to reduce the complexity of the interface matching algorithm and increase its accuracy. The proposed approaches for interface matching adapt schema matching techniques, by using WSDL information (messages associated to operations, message direction) and also by encoding the protocol information in terms of depth and propagation of similarities to neighbor messages (incorporating the notion that messages with similar depth in the two protocols P1 and P2 are more likely to match) and also disallowing matchings leading to deadlock cases.

In [474], authors give a formal semantics to protocol matching based on FSM extended by logical expressions attached to states. These annotations, expressed over the outgoing transitions of states, specify the mandatory transitions of a state that a potential FSM match must support. For instance, let us consider the FSM depicted in Fig. 3.1a whose state 2 is annotated with *GetAddress and GetZone*. This annotation means that the potential matcher of this process must have a state supporting both transitions *GetAddress* and *GetZone*. In this work, authors reduce the process matching problem to automata intersection problem, which is expressed as finding shared traces between two annotated FSMs. Therefore, two annotated FSMs match if they share at least one common path between the start and end states, and all the mandatory transitions specified by the query process are supported by the target one. For instance, the process depicted in Fig. 3.1a does not match the processes depicted in Fig. 3.1b (because it does not support both transitions *GetAddress* and *GetCountry*) and Fig. 3.1c (because they have no common trace), while the process depicted in Fig. 3.1d is a match since it satisfies mandatory annotation (it supports both transitions *GetAddress* and *GetCountry*) and shares two traces with this process. It should be noted that even though the set of traces can be

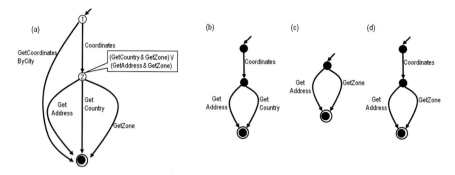

Fig. 3.1 Example of annotated FSM intersection-based matching

derived from an FSM, this can be potentially infinite or at least exponential in size of the FSM.

To assess similarities and differences between two business protocols specified using FSMs, as well as whether they can interact or not with each other, a set of algebraic operators is defined in [68], which are:

- Compatible composition operator ($||^C$) computes an FSM characterizing possible interactions between two business processes.
- Intersection operator ($||^I$) computes an FSM that describes the largest set of execution traces that are shared between the two FSMs given in parameters.
- Difference operator ($||^D$) computes the set of traces of a given FSM that are not shared with the second one.
- Projection operator ($[P_1||^C P_2]_{P_1}$) computes the part of P_1 (also as an FSM) that is involved in the compatible composition of P_1 and P_2.

These operators analyze the processes in light of their branching-time semantics [452], where possible execution sequences allowed by a process are characterized in terms of execution trees [222], instead of execution paths.

The problem of protocol compatibility is addressed from a different perspective in [254], by representing the set of all behaviorally compatible partners to a given service, called *operating guidelines*. This set is represented using the concept of annotated automata and can be used for applications like substitutability of services and contract-based composition. The basic set operations union, intersection, and complement and membership and emptiness tests are implemented on an extension of annotated automata (representation of sets of services). The relevance of these operations is motivated in three applications. For service substitutability applications, one of the advantages of the proposed approach is that it can return counterexamples in case of nonsubstitutability.

In the same vein, authors in [353] propose two operators to manage processes captured as hierarchical statecharts,[3] which are: (a) Match, for finding correspondences between models, and (b) Merge, for merging processes with respect to known alignment between them. *Match* operator takes into account the syntactic and linguistic similarities of state names and behavioral similarity of states using approximations of bisimulation [334] (the similarity between two states depends on the similarity of their successors and predecessors), as well as the depths of states in the statechart hierarchy trees. Because of the use of state depth to measure the similarity of states, this technique can be applied only on hierarchical statecharts. *Merge* operator takes the identified mappings between two statecharts and computes the merge process that better preserves the behavioral constraints specified by the two statecharts.

Many other notions of behavioral equivalence (e.g., bisimulation, branching bisimulation, etc.) can be used to compare process models (see [383] for an overview of equivalence notions in the context of Petri nets). However, most of the existing approaches are not very appropriate for process matching since they result in a binary answer (i.e., two processes are equivalent or not). Moreover, such kinds of techniques are mainly based on state-space search, which is computationally expensive.

3.5 Process Matching

Providing an effective solution to the problem of process model matching requires handling three process model aspects, which are terminology, granularity levels, and behavioral/structural differences. Process models (whose activities are semantically annotated) depicted by Fig. 3.2 are used to exemplify these aspects. The terminology mismatches are related to the differences between metadata (name, inputs, and outputs) describing activities. For instance, activity *GetBranchNumber* in *T* is a match of the activity *GetCABOfBank* in *Q*. Even if their names are syntactically different and their outputs are not the same (they have the same super concept), the algorithm must be able to match them approximately. In other words, the algorithms must be able to match activities even if they are described differently. The second difference is related to the differences in granularity levels of *Q* and *T*, where the activity *GetBranchInfo* in *T* is matched against the sequence of target activities *GetBank* and *GetBranchInfoByBankName* in *Q*. Composition techniques that allow detecting such kind of matches have to be proposed. The third aspect concerns behavioral differences and/or similarities. For instance, *GetCABOfBank* and *GetABIOfBank* activities of *T* are in sequence, while their activity matches in *Q* are in parallel. The algorithm must match similar behaviors even if they are specified using different control constructs. Furthermore, the algorithm would be

---

[3]Statecharts is an enhanced realization of the finite state machine [217].

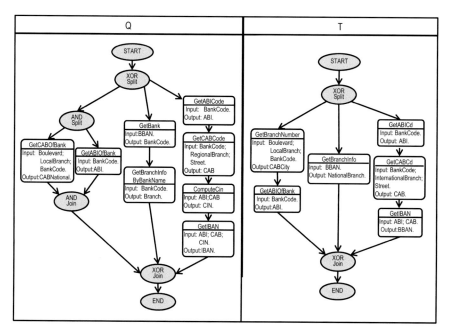

Fig. 3.2 Process model mismatches

able to consider user behavioral similarities/differences definitions to check whether two control constructs are similar or not since no consensus is made on how two behaviors are similar.

Measuring the similarity of two process models is approached in the literature by considering the structural or behavioral perspectives. Behavioral-based matching approaches relate to the process execution semantics by comparing the traces generated by the two processes. A trace is an execution of a process instance starting from an initial state and ending into a final state that consumes a set of inputs and produces a set of outputs. Structure-based matching approaches consider only the process topologies, i.e., the type of the control flow specified in the process models. To illustrate the difference between the two approaches, let us consider the two processes PM_1 and PM_2, depicted by Fig. 3.3. From the behavioral perspective, they are similar since they have the same set of traces, i.e., the two processes start by send/receive message m_1, possibly followed by a sequence of messages m_2 and m_1 and ended by send/receive messages m_3 or m_4. In contrast, PM_1 and PM_2 are different from the structural perspective since the initial state of PM_1 has an outgoing transition labeled with m_1 and ingoing transition labeled with m_2, while the initial state of PM_2 has just an outgoing transition labeled with m_1.

In the following, we present an overview of the techniques addressing the problem of process model matching. We first present different formalisms for representing process behavior or structure. Then we present techniques for mapping

Fig. 3.3 Structural vs.
behavioral matching

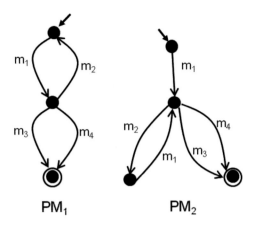

PM_1 PM_2

process activities, which is the first step of process comparison. Then we present
techniques for matching process behavior and structure.

3.5.1 Abstract Models for Process Model Descriptions

Current approaches for matching processes operate on different abstract formalisms
that represent processes. In the following, we present the most often used formalisms, which are finite state machines (FSMs), process graphs, and workflow
nets.

3.5.1.1 Finite State Machine

A FSM [22] is a model for process description composed of a finite number of states,
transitions between these states, and messages. Mathematically it is represented as a
tuple $P = (S, s_0, F, M, R)$, where, S is a finite set of states, $s_0 \in S$ is the initial state,
$F \subseteq S$ is a set of final states, M is a finite set of messages, and $R \subseteq S^2 \times M$ is a finite
set of transitions. Each transition (s, s', m) identifies a source state s, a target state
s', and a transition label m representing the message sent or received by the process.
States represent the different phases a process go through during its execution,
and transitions represent the shift from one state to another by sending/receiving
a message that is used as a transition label.

FSMs are able to capture the sequence, alternative, parallel, and loop control flow
structures offered by process specification languages. A conceptual model based on
FSMs is presented in [69] that allows capturing a large part of process semantics
and temporal constraints.

However, FSMs in their original form do not provide means to model concurrency control flow structure. Thus, all the sequences induced by a concurrent

execution of a set of activities should be enumerated to capture such control structure (see [319] for an algorithm enumerating these sequences). Thus, they are more frequently used to model simple behavior, like business protocols.

FSMs offer a graphical notation for stepwise processes and have a formal semantics, with a well-developed mathematical theory necessary for analyzing certain process properties, such as liveness (i.e., a process is ensured to move toward a state where a goal can be reached) and safety (i.e., at each state of the FSM a logical invariants hold).

3.5.1.2 Process Graphs

Graphs are a very common formalism for modeling temporal and logical dependencies in processes. Graphs are the basis of several process specification languages such as event-driven process chain (EPC for short) [326] and business process modeling notation (BPMN for short) [365].

A process graph [469] is a tuple $PG = (A, C, L, \varphi, \lambda, E)$ consisting of two disjoint and finite sets: activities (A) and connectors (C); a set of activity metadata L; a mapping $\varphi: A \to L$ which is a function that maps activity nodes to their metadata; a mapping $\lambda: C \to \{and\text{-}split/join, or\text{-}split/join, xor\text{-}split/join\}$ specifying the type of a connector $c \in C$ as AND (parallel execution of branches), OR (alternative execution of branches), or XOR (exclusive alternative and loop execution of branches); and a binary relation $E \subset (A \cup C)^2$ defining the control flow constraints between nodes (edges).

Process graphs are able to capture the major control flow structures offered by the process specification languages (sequence, alternative, concurrence, and loops). In [327], a technique enabling transforming block-oriented process specification languages (e.g., BPEL [161], OWL-S) to EPCs is presented.

A major strength of process graphs is their simplicity and easy-to-understand graphical notation. They are used by several process similarity evaluation approaches [137, 302, 336, 451] as a formal representation of processes.

3.5.1.3 WF-Nets

A workflow net (WF-Net for short) [379] is a special class of Petri net used for modeling inter-organizational workflows. A WF-Net is a directed bipartite graph represented as a tuple $W = (P, T, F, L)$ in which T represents a set of transitions, P is a set of places, $F \subseteq (T \times P) \cup (P \times T)$ is a set of directed arcs, and L is the labeling function which assigns a label to each transition. Transition labels represent the messages that the workflow sends or receives. Input arcs connect places with transitions, while output arcs start at a transition and end at a place. Places contain tokens, and the number of tokens in each place represents the current state of the workflow, called the marking. The workflow marking changes when transitions are ready to fire, meaning that there are enough tokens in the input places of these

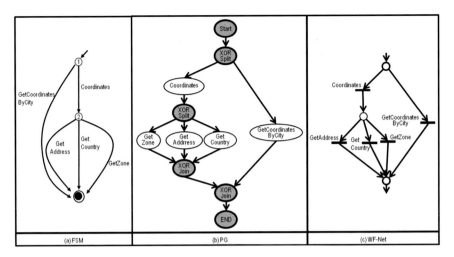

Fig. 3.4 Example of a process specification using (**a**) FSM, (**b**) PG, and (**c**) WF-Nets

transitions. A WF-Net must have one initial place and one final place, and every transition and place should be located on a path from the initial and the final places.

WF-Nets offer a graphical notation for stepwise processes that allows capturing sequence, alternative, parallel, and loop control flow structures. WF-Nets provide an exact mathematical definition of their execution semantics, with a well-developed mathematical theory for process analysis (liveness, reachability, deadlock free, etc.).

An illustration of these abstract models is presented by Example 3.1.

Example 3.1 Figure 3.4 shows a *geographical location* process specified using the abstract models described above. The process retrieves the address, the country, or the zone of geographic coordinates. It also retrieves the geographic coordinates of a city.

3.5.2 Matching Process Activities

When comparing two processes, first a mapping between activities of the two processes has to be found. Techniques described in Sect. 3.3.1 for matching operation names can be applied to evaluate similarity of activity names. As we explained previously, activity labels used in process models are often formed by a combination of words and can contain abbreviations. To obtain a linguistic distance between two strings, different string matching algorithms can be used: tokenization, n-gram [27], Check synonym, Check abbreviation, tokenization, stemming, etc.

Several tools have been developed recently [103] to find the best mapping between the sets of activities of two processes modeled as Petri nets. Triple-S, a tool based on the approach presented in [157], firstly performs two preprocessing steps: tokenization and stop word elimination. The similarity between two activities is based on the Levenshtein distance [287] and the Wu and Palmer similarity [475] between tokens (using the WordNet lexical database) and structural aspects: the ratio of their in- and outgoing arcs and their relative position in the process.

RefMod-Mine/NSCM [103] adopts a different approach, based on a clustering method. The nodes of all models are being compared using a similarity measure based on the Porter stem sets [386]. A preprocessing phase aims at ignoring wrong modeled transitions and a post-processing phase check labels for antonyms using WordNet and looks for negation words (like not). The cluster algorithm is agglomerative; it starts with clusters of size 1 and then consolidates two activities to a cluster if their similarity passes a user-defined threshold. Given a pair of process models, the last phase of the algorithm extracts a complex (N:M) matching by scanning clusters for the occurrence of nodes of both models.

The "Bag-of-Words Similarity" tool is based on the approach presented in [263] and treats each label as a bag of words (a multi-set of words). It applies word stemming and then computes similarity scores using Levenshtein and Lin [298] similarity measures for each pair of words. Furthermore, in case the two bags of words are different in size, the larger one is pruned by removing words with a small similarity score. Finally, an overall matching score is calculated, and all activity pairs whose score is above a given threshold are selected.

A more difficult problem is finding complex mappings, when an activity in one process represents a functionality that was implemented as two or more activities in the other process. After identifying that one activity matches several activities in the other process, this task requires also to find clues that these activities should be composed to match the unique activity. In [197], complex mappings are retrieved between a synchronous and two asynchronous activities that implement the same functionality. In [177], complex mappings (1-n) between activities are found based on their inputs/outputs allowing to identify granularity differences in implementing a given functionality. Figure 3.5 shows an example of such granularity differences. The query process specifies the functionality for updating user contact

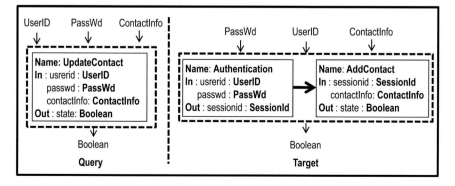

Fig. 3.5 Granularity-level differences

information as a single activity (*UpdateContact*), while the target one implements it as a sequence of two activities (*Authentication* followed by *AddContact*). Since *Authentication* and *AddContact* query activities are in sequence and the union of their inputs and their outputs correspond, respectively, to the inputs and outputs of UpdateContact target activity, then a complex mapping is identified between the composition of *Authentication* and *AddContact* and *UpdateContact*.

Weidlich et al. [466] propose the ICoP framework for identifying complex correspondences between groups of activities. The framework has a set of reusable matchers and allows to develop new matchers from the existing ones. Dhamankar et al. [136] present an approach for semiautomatically identifying complex matches between data sources, such as **address**=concat(**city, state**). To improve matching accuracy, the approach uses different types of domain knowledge, such as past complex matches and domain integrity constraints. Although their approach has been proposed in schema matching, it could be applied to other data representations (e.g., graph data structures).

Taking into account 1-to-N or N-to-M correspondences is required also in the setting of loosely coupled and distributed process execution data. In [42], the authors propose an approach that handles N-to-M relations between events (logged by IT systems) and activities. To identify such relations, they exploit domain knowledge derived from process documentation.

3.5.3 Structural Matching

Structure-based approaches are in general graph based and take into account either the global graph topology or local structural relationships between activities.

In [197], authors reduced the problem of process model matching to a graph matching problem, and an error-correcting subgraph isomorphism (ECSI for short) detection algorithm [137] was adapted to this purpose. The principle behind ECSI algorithm is to apply edit operations (node/edge insertions, deletions, and substitutions) over the target graph until there exists a subgraph isomorphism to the query graph. Each edit operation is assigned a cost function, on the basis of which the quality of an ECSI is estimated. The goal is then to find the sequence of edit operations leading to the ECSI between the query and target graphs that has the minimal cost. To find this sequence, the ECSI detection algorithm relies on an exhaustive A* search space algorithm [419], where each state represents a partial mapping between the nodes of the given graphs, and a transition between two state s_1 and s_2 represents the extension of the mapping represented by s_1 by adding a pair of mapped nodes. Each partial mapping implies a number of edit operations whose cost is used to evaluate the quality of the state representing that mapping. For instance, the cost of substituting two activity nodes depends on the similarity of their names, inputs, and outputs. The graph representation of a BPEL process model has two kinds of nodes: activity nodes and connector nodes (expressing synchronization constraints between two activities). The matching algorithm compares also the connectors (considered as edges), which allows detecting differences of the control flows (synchronization constraints), for example, a sequence of two activities could be matched against two activities connected in parallel.

Another major feature proposed by the authors concerns the detection of granularity-level differences between the process graphs to be matched, i.e., the matching algorithm is able to match a single activity in one graph against a sequence of two activities in the other graph. This is done by introducing new edit operation: merging edit operation that allows composing many activities and splitting operation that allows dividing an activity into a sequence of activities. A method generalizing this principle and allowing matching a block of activities in one graph against a block of activities in the other graph is proposed and validated over OWL-S process models [177].

Experimental results showed their effectiveness in terms of precision and recall for process matchmaking and ranking, but the underlying graph matching algorithms are exponential in size of the graphs, which limits their application in practice to graphs of small sizes. Two heuristics are proposed in order to overcome this issue. The idea behind the first heuristic (called input/output driven matching algorithm[176]) is to consider only the pairs of activity matches respecting an acceptable similarity level when exploring the search space of the ECSI detection algorithm. The goal is then to establish, in a pre-computation step, the set of activity match candidates that the ECSI detection algorithm can explore. These match candidates are established on the basis of the matches found between the profiles (the sets of inputs and outputs of the processes) of the two processes to be matched, by considering that two activities match if they share at least one input or output. Thus, for each activity of the target process, a set of its match candidates in the query process, including 1-1 and 1-many matches, is established. This step of searching activity match candidate guarantees the exploration of only the promising branches of the ECSI detection search space, which allows reducing significantly the execution times. The idea of the second heuristic (called summary-based matching algorithm [178]) is to reduce the number of nodes of the processes to be matched by exploiting the fact that processes are organized into blocks representing particular phases of their executions. Consequently, the approach transforms these blocks into summary nodes. This transformation leads to summary processes having smaller sizes, and thereby their matching requires less execution time. A set of summarization rules are defined that ensures the uniqueness of the summary of a given process.

These summary processes are then matched using an adapted version of the ECSI detection algorithm, following the intuition that similar activities are grouped into similar summary nodes. An extensive experiment evaluation of these heuristics is presented in [175]. In [137], authors adapt also the ECSI detection algorithm and propose three heuristics to improve its scalability. The first one is a greedy heuristic, in which only one mapping is explored. This mapping is created and extended at each iteration, by adding the pair of the most similar nodes that are not yet mapped. Unlike the greedy heuristic, the other two heuristics explore several possible mappings (like A* algorithms) but use pruning techniques that tailor the search space (only the most promising mappings are explored). The first technique prunes the search space as soon as the number of explored mappings exceeds a certain threshold and keeps only the mappings with the highest similarity. The second pruning technique firstly starts by mapping the source nodes of the process models, then maps the nodes that immediately follow the source nodes, and so on. The latter may, in some cases, allow the algorithm to converge rapidly toward the optimal solution. The experimental results show that the greedy heuristic presents a good trade-off between the computational complexity and the matching quality.

The graph-edit distance technique was also used in [336] to measure, in the context of the agile management of business processes [464], the similarity between two changed/adapted process instances of the same process. The distance is viewed as being the number of activity nodes and control flow constraints that are not shared by the two workflow instances. The distance between two processes P_1 and P_2 is expressed by the following formula: $Dist(P_1, P_2) = |\hat{N_1}| + |\hat{E_1}| + |\hat{N_2}| + |\hat{E_2}|$, where $\hat{N_1}$ (resp. $\hat{N_2}$) are nodes within P_1 (resp. P_2) but not within P_2 (resp. P_1) and $\hat{E_1}$ (resp. $\hat{E_2}$) are edges within P_1 (resp. P_2) but not within P_2 (resp. P_1).

Instead of comparing two processes in terms of the number of atomic edit operations (adding/removing nodes and edges) needed for transforming one process to another (like [137, 336]), authors in [294] propose an edit distance based on high-level change operations (formalized in earlier work [394]) necessary to achieve this transformation. These high-level change operations, defined on the basis of atomic edit operations, include inserting, deleting, and moving an activity, parallelizing a sequence of activities, sequencing parallel activities, etc. In this approach, the atomic differences between processes are first grouped using patterns defining the high-level change operations, and a similarity is computed on the basis of these grouped differences. This approach provides a more meaningful measure of the efforts needed to transform a process to another one.

In [302], authors present a technique for measuring the similarity degree of two process variants (P_1 and P_2) of the same process, in terms of three structural matching levels that are (1) equivalent match when P_1 and P_2 have the same activities and the same control flow constraints; (2) subsume match when P_2 nodes belong to P_1 nodes, and P_2 preserves the structural constraints specified in P_1; and (3) imply match when P_1 and P_2 have the same activities, and P_2 preserves the structural constraints specified in P_1. The technique first eliminates from P_1 all the activities that are not contained in P_2, and then it reduces the redundant flow relations in P_1 using a set of reduction rules defined in earlier work [408]. P_1 and P_2 match if the reduced graph of P_1 holds the "equivalence" or "subsume" relationship with P_1.

In [157], process models formalized using Petri nets are transformed OWL ontology, and the method designed for aligning ontology was adapted to match the transformed process models. The similarity is computed at the basis of the similarity of the ontology element names (places, transitions) which is calculated by aggregating syntactic, linguistic, and structural similarity. Syntactic similarity is based on Levenhstein distance [287]. The linguistic similarity degree between process element names is based on the synonym sets proposed by Wordnet: if the two names have a common sense (the intersection of their senses sets is not null), their similarity is 1 divided by the maximum sense cardinality of the two terms (otherwise, the similarity is 0). The structural similarity measure has as input two concept instances and their contexts which is defined as the set of all elements which influence the similarity of the term. The context for place names include attributes (workflow variables) and successor transition names. The similarity of attributes is influenced by their values and their siblings. The context for transition names include all its successor place names.

In [259], authors investigate the use of iSPARQL framework [260] for semantic process retrieval. iSPARQL extends SPARQL [160] with similarity operators to be able to query for similar entities in semantic Web knowledge bases. SPARQL is an emerging standard query language for RDF. The framework was used to query the OWL MIT Process Handbook, which contains a large collection of semantic business processes. Queries allow to find processes which are similar to a given process by comparing process names and descriptions and specifying different similarity strategies. Three simple and two human-engineered similarity strategies were evaluated. Moreover, machine-learning techniques were applied to learn similarity measures showing that complementary information contained in the different notions of similarity strategies provides a very high retrieval accuracy.

The authors of [431] present some preliminary ideas for taking into account process specifications for semantic Web service discovery. They propose to represent the query as a collection of relation expressions (similar to an RDF statement $< subject, property, object >$) for preconditions, effects, guards, and processes. They envisage to extract the rule expressions embedded in the process specification, translate them into a rule language, and use a rule reasoning engine to check whether the process satisfies the query. After matching functional requirements, process flow constraints are verified by associating each service in the process with guards (conjunction of all the branching conditions on the path the service belongs). To check whether a service is executed under the context of a particular query, guards are evaluated using the information provided in the query.

3.5.4 Behavioral Matching

In this section, we show how execution semantics of the processes is used, in the literature, to achieve process similarity evaluation.

Considering the fact that some traces of a process are rarely triggered while others are executed for the most cases, authors in [323] propose a similarity metric for evaluating the closeness of two processes (specified using WF-Nets), which integrates the traces and their occurrence probabilities, i.e., differences between rarely triggered traces of processes are less important than differences in the frequently triggered traces. The defined measure is computed at the basis of an event log containing trace occurrence probabilities that can be obtained from process execution logs, from user-defined scenarios, or by simply simulating the models.

To overcome the performance problems, due to large sets of generated traces, potentially infinite, when using the behavioral perspective of processes to measure their similarities, approximations of the behavior of a process are proposed in the literature, which are causal footprint [451] and process type [161]. A causal footprint [451] is defined as a tuple $P = (N, L_{lb}, L_{la})$, where N is a finite set of activities, $L_{lb} \subseteq (P(N) \times N)$ is a set of look-back links, and $L_{la} \subseteq (N \times P(N))$ is a set of look-ahead links. Each look-ahead link $(a, B) \in L_{la}$ means that the execution of a is followed at least by the execution of an element $b \in B$. Each look-back link in $(A, b) \in L_{lb}$ means that the execution of b is preceded at least by the execution of an element $a \in A$. Causal footprints of processes are then represented in a document vector space model whose dimensions are formed by activities, look-ahead links, and look-back links that belong to at least one causal footprint representing one process. The distance between two causal footprints is then the cosine of the angle between the vectors representing them.

A process type [161] captures the essential behavioral characteristics of a process by representing it as a set of behavioral relations that may occur between its activities. The behavioral relation between two activities may be one of the following:

- SEQ relation: $Act_1 <_{SEQ} Act_2$ means that Act_1 is executed before Act_2.
- XOR relation: $Act_1 <_{XOR} Act_2$ means that Act_1 and Act_2 are never executed in the same trace.
- AND relation: $Act_1 <_{AND} Act_2$ means that Act_1 and Act_2 are executed in parallel.

The similarity between two processes is defined as the ratio between the size of the behavioral relations shared between the two processes and the size of the largest process type. The approach is exemplified using BPEL4WS [30] processes but it could be applied, in principle, to other processes.

This similarity measure is generalized in [276] for any process description language using the notion of behavioral profile. Similarly to process types, behavior profiles contain also the set of relations between activity pairs, but they are calculated based on the observed order of activities in traces, and not on the BPEL model. Based on these relations, five elementary similarity measures are calculated using the Jaccard coefficient. Finally, these measures are aggregated using a weighted mean. The proposed behavioral profile metric is proven also to be a metric, that is, to have the properties of symmetry, nonnegativity, identity, and triangle inequality. The authors show that the triangle inequality allows an efficient similarity search in large process collections, by using metric trees [275]. Metric trees can be used as an index to partition the collection of process models, certain partitions being pruned during search and thus avoiding the comparison of the query model with all the models.

Zihe Dong et al. [147] and Zixuan Wang et al. [461] propose behavioral similarity measures for WF-Nets based on the coverability tree. The coverability tree of a Petri net is generated by iterating all the markings that can be reached; it has markings as nodes and the firings of transitions as edges. The CFS algorithm [147] first constructs the set of complete firing sequences on coverability tree (including loop identification and count). Then, model similarity is calculated by applying the A* algorithm to construct an optimal match between two sets of complete firing sequences. Two complete firing sequences are compared based on their longest common subsequence. The second approach, called TAGER [461], transforms the coverability graph of a Petri net into its isomorphic graph, named T-labeled graph, by moving the labels of transitions to their successive vertices. The goal is to have transition names associated to vertices, to facilitate the computing of the similarity measure. The similarity of the two T-labeled graphs is calculated based on the graph-edit distance using a cost function adapted for this type of graph.

3.6   Discussion and Concluding Remarks

As we have seen in this chapter, when comparing process models, there is a need to consider different types of attributes of process models, such as labels, textual description, semantic description, and graph-based models. For the similarity computations, we can rely on a variety of similarity functions, such as string matching functions (e.g., edit distance and its variations, Jaccard similarity, and tf-idf-based cosine functions), numeric matching functions (e.g., Hamming distance and relative distance), structure matching functions (e.g., graph-edit distance and similarity flooding), and so on. However, no single similarity function can perform well for all different types of perspectives. It is a hard decision to determine which similarity functions should be used for given matching tasks as:

- Choosing appropriate similarity functions is highly domain and data dependent. Even a function that shows good performance for some datasets can perform poorly on new and different ones.
- Selecting appropriate functions could require a high manual effort as well as lots of experience and expert knowledge, due to the large number of available similarity functions and the wide variety of data characteristics.

In [406, 407], the authors propose an approach for recommending which similarity functions (e.g., edit distance or Jaccard similarity) should be used for a given similarity search task. The approach employs an incremental knowledge acquisition technique for capturing domain experts' knowledge about similarity functions and their usage contexts. In addition, for situations where domain experts have little or no knowledge about datasets (e.g., when the experts face *new* or *different* ones), they analyze data features (e.g., misspellings or word orders), which are considerable when selecting similarity functions so that similarity functions are recommended based on the identified features.

Once an appropriate similarity measure is chosen for each atomic attribute, the next step is to aggregate them. We have seen in the previous sections that many process matching techniques have been developed, most of them addressing one dimension of process description: interface, structure, or QoS properties. An exception is the work presented in [13, 286] which addresses both structural similarity

and QoS satisfaction degree. Building a complete solution addressing all the aspects
raises several challenges. First, choosing the order in which the algorithms matching
particular process attributes should be applied. As shown in [13], sometimes they
should be interwoven to reduce the search space or to improve the precision of the
search methods. Second, how to combine the similarity measures for the different
perspectives? One can use the weighted average, personalized aggregation metrics
like fuzzy linguistic quantifiers (such as almost all) or the skyline operator. A third
question is which technique to choose for each process perspective. The choice
of the interface matchmaking depends on the richness of the interface description:
presence of textual description, semantic annotations, etc. Matching QoS constraints
is easier when a common QoS ontology or vocabulary is used.

Table 3.1 summarizes the characteristics of the matching techniques presented in
the previous sections for comparing business protocols or process models. They are
compared according to the following criteria:

1. Abstract model criterion classifies approaches according to the abstract model on
   which the matching algorithm operates.

Table 3.1 Table summarizing matching algorithm characteristics

	References	Abstract model	Match type	Theorical complexity	Activity similarity
Structural	[197]	Graph	Exact, inexact	NP-Complete	String distance + WordNet
	[176–178]	Graph	Exact, inexact	NP-Complete	String distance + ontological similarities
	[137]	Graph	Exact, inexact	NP-Complete	String distance
	[336]	Graph	Exact, inexact	NP-Complete	Name equality
	[302]	Graph	Exact	Not given	Name equality
	[294]	Graph	Exact, inexact	Not given	Name equality
	[157]	Petri net	Exact, inexact	NP-Complete	String distance + WordNet
Behavioral	[474]	FSM	Exact, partial	NP-Complete	Name equality
	[68]	FSM	Exact, partial	Polynomial	Name equality
	[353]	Statechart	Exact, inexact	Polynomial	N-gram
	[451]	Graph	Exact, inexact	Not given	String distance + WordNet
	[323]	WF-Nets + Log	Exact, inexact	Not given	Name equality
	[161]	Tree	Exact, inexact	Not given	Name equality
	[276]	Graph	Exact, inexact	Cubic for sound free-choice WF-net	Name equality
	[147]	WF-Nets	Inexact	Not given	Name equality
	[461]	WF-Nets	Inexact	Cubic	Name equality

2. Match type criterion compares approaches in terms of the type of the found matches. Three types of matches can be considered: exact, inexact, and partial. Two processes match exactly if they have the same structure or behavior. Partial match refers to two processes that share some traces. The inexact match refers to the fact that two processes are not identical, but they are similar at some degree.
3. Theorical complexity criterion provides the class of complexity of the approaches (if given in the original paper).
4. Activity similarity criterion recalls the techniques used for assessing the similarity of activities.

The table is divided into two parts, according to the process perspective (structure or behavior) considered by the techniques.

The majority of techniques supporting inexact matches are NP-complete (the search space is fully/partially explored). Despite efforts to make the algorithms tractable via applying heuristics, their applications in practical cases are limited generally to small-sized processes [137].

The approaches based on the approximations of the behavior of processes (process types [161] and causal footprint [451]) achieve good execution times. However, because the similarity between two processes is based on the number of relations shared between them, their accuracy is not sufficient (two processes can share behavioral relationships, while they can be behaviorally different) for the applications requiring 1-to-1 process matching algorithms. Knowing their performances in terms of execution time, they can be recommended as filtering steps (they give an overall indication of the similarity of two processes) before applying other algorithms for more accurate matches. Exception is the matching algorithm designed for matching statecharts [353] which can be done in a polynomial time. However, the aim of this work is to compute the correspondences between the activities of two statecharts in order to merge them; no similarity metric is defined to measure the overall similarity of two processes.

As we can see it in Table 3.1, the majority of the proposed techniques consider that the processes to be compared use a shared vocabulary. Thus, in order to apply these proposals for processes that were developed independently and do not use a shared vocabulary, efficient techniques allowing finding mappings between activities have to be applied in a preprocessing step.

A survey of existing process similarity measures is presented in [50]. The authors elaborate a number of desirable theoretical properties for similarity measures, and they analyze 23 similarity measures with respect to these properties. The measures are also used to calculate the similarity between an example model and its variations. The results show that similarity measures rank processes very differently and that, in general, they do not fulfill all desirable properties. As a conclusion, the authors give some recommendation for the selection of an appropriate similarity measures for different applications.

A different approach for selecting a similarity measure is presented in [468]. The proposed method allows to predict the quality of results derived by process model matchers. The prediction is intended to be used as a decision-making tool about the trust one should put in automatic matching.

This chapter presented techniques for process matching. We overview similarity measures that can be considered for different types of process attributes. Then we presented techniques that compare process models from one of the main perspectives: interface, business protocol, or process model. Depending on the application, the appropriate technique or a combination of techniques could be used. As we have seen in Sect. 3.1.1, process matching techniques are useful for managing process model repositories (for implementing model operators like match, merge) and for process similarity search. The later application aims at finding in a process model repository, the models similar to the user one. An existing process model is used as a query. In other applications, users do not have a complete process model for specifying their needs, but they are interested to find process models having particular characteristics or satisfying some constraints. For example, when analyzing business process models for checking compliance rules, a user may want to find answers to questions like the following: *are there any process models (or instances) where activities A and B are successive?* Query languages have been proposed for allowing to find processes containing patterns of interest. The next chapter will present such query languages for querying process model repositories or process execution data.

Chapter 4
Model-Based Business Process Query Techniques and Languages

Nowadays, it has become increasingly common for organizations to work in a process-oriented manner. As discussed in the previous chapter, single organizations may be dealing with collections of hundreds or thousands of business process models. Examples of such collections include the BIT process library [165], the SAP reference model [126], the Dutch municipalities[1], and the Suncorp's process model repository [281]. In general, business process models are created by business analysts with the objective of capturing business requirements, enabling a better understanding of business processes, facilitating communication between business analysts and IT experts, identifying process improvement options, and serving as a basis for derivation of executable business processes. Business processes are executed with the objective of generating value to the organization's customers, be it external or internal customers. However, business processes must not be designed in any arbitrary way, as the organization has to adhere to governing standards, laws, and other regulations. As organizations develop large repositories of business process models, systematic access to these artifacts is essential, and business process model querying techniques are crucially required to ease and unify access to the artifacts and knowledge of these process models.

In practice, business process models are not built just for communication purposes among business experts. Processes are executable. In addition to proprietary workflow languages, standardized executable processes exist. For example, Business Process Execution Language for Web Services (BPEL4WS) is a standardized executable language [125] with major software vendors supporting it. In principle, executable process languages orchestrate invocations of individual applications either internal or external to the organizations. There are many scenarios where the execution of these processes needs to be monitored and analyzed. Therefore, querying techniques for the business process executions serve as fundamental components to realize the monitoring and analytics goals.

---

[1]http://www.model-dsp.nl/

© Springer International Publishing Switzerland 2016
S.-M.-R. Beheshti et al., *Process Analytics*, DOI 10.1007/978-3-319-25037-3_4

The goal of this chapter is to discuss and provide an overview of the various techniques for querying business processes. In particular, Sect. 4.1 provides an overview of the different techniques for querying repositories of business process models, while Sect. 4.2 discusses the different techniques for querying the business process execution logs and their related artifacts. Section 4.3 provides an overview of different approaches for utilizing business process querying techniques for ensuring business process compliance to their specifications and business rules.

4.1 Querying Repositories of Business Process Models

To understand, communicate upon, or reengineer working procedures, companies document their daily routines in the form of business process models. Therefore, business process models represent the blueprint for subsequent executable processes. In general, business process modeling is considered to be a complex, time-consuming, and error-prone task. Model design requires determining the activities that need to be performed, ordering of their execution, handling exceptional cases that might occur, etc. In addition, in many cases, variants of process models need to be created in response to special business situations. For instance, there could be several insurance claim handling processes. One process is designed for people with special working environment conditions, another for people over 70 years, etc.

In practice, querying for a specific process model or a model fragment within a repository of business process models has multiple uses. For example, given a collection of business process models, querying can be used to retrieve models that have specific properties, such as a specific activity or a specific relation among activities that they contain. In addition, it can be used to identify process models that do or do not comply with given standards or internal practices, process models that are candidate for refactoring, or process models that can be reused as a template to build new ones [139]. In principle, providing business process designers with querying mechanisms that enable reusing previously designed business process models can significantly simplify and improve the business process modeling task and effectively increase the quality and the maturity of the newly developed process models.

Various approaches and query languages have been proposed to express and execute queries over repositories of business process models such as *BPMN-Q* [36, 410], *BP-QL* [51], *BeehiveZ* [242, 244], and *APQL* [439]. In general, these languages use a declarative approach where the user can express his/her query specifying the existence or absence of specific elements on the business process models. In addition, queries can be formulated over control-flow aspects of process models such as direct connections or transitive paths. However, these languages vary with respect to their expressive power. For example, some languages may embrace some special aspects such as the resources that are used to perform the process, the process trigger, the process goal, the location(s) at which the process is executed, or the categories in which the process is classified. In addition, these approaches vary in their query evaluation mechanism. For example, some approaches translate the input query into a language that can be understood by the underlying technology which is used to store the business process models. For instance, if the underlying storage technology is a relational database or an XML database, then the business process query languages are typically transformed into SQL [36, 410] or XML query languages [51, 114], respectively. Some other querying approaches employ subgraph isomorphism search algorithms [242, 244] where indexes can be used to speed up the query execution using the commonly used filtering-and-verification strategy on handling graph databases. In the following sections, we give an overview and discuss the different design decisions for each of these approaches.

4.1.1 BPMN-Q

BPMN-Q [36, 410] is a visual business process query language that relies on the notations of BPMN languages as its concrete syntax. Therefore, it shares many visual notations with the BPMN modeling language. Moreover, it provides a set of new constructs that can be seen as abstractions over the existing modeling constructs. In principle, BPMN-Q is used to query business process models by matching a process model graph to a query graph. Figure 4.1 illustrates the main visual BPMN-Q querying constructs which are described as follows:

- **Path edges:** a path edge connecting two nodes in a query represents an abstraction over whatever nodes could be in between in the matching process model.
- **Undirected data flow edges:** this type of edges is used to connect a data object to a path edge in a query. This is also an abstraction mechanism where we look for paths on which there are nodes that access the specified data object.

Variable Node	It is used to indicate unknown activities in a query. It resembles an activity but is distinguished by the @ sign in the beginning of the label.	@Variable
Generic Node	It indicates an unknown node in a process. It could evaluate to any node type.	*
Generic Split	It refers to any type of split gateways.	S
Generic Join	It refers to any type of join gateways.	J
Negative Sequence Flow	It states that two nodes A and B are not directly related by sequence flow.	—x→
Path	It states that there must be a path from A to B. A query usually returns all paths.	—//→
Negative Path	It states that there is not any path between two nodes A and B.	—x//→

Fig. 4.1 Basic BPMN-Q constructs

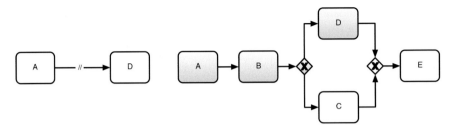

Fig. 4.2 An example BPMN-Q query with a match to a process model

- **Anonymous activities and data objects:** they are used by users who may need to issue a query, what activities read/update the insurance claim data object, on the form. Since the user does not know that activity, he/she can start its label with the "@" symbol to declare it as an anonymous activity the query processor has to find. Similarly, data objects can be anonymous.
- **Negative control-flow and path edges:** the user is able to express in a query that nodes A and B must have no control-flow edge from A to B (negative control flow), or there must be no path at all from A to B (negative path).
- **Generic ControlObject and GateWay:** these classes are no longer *abstract*, i.e., in a query, the user is able to put a generic node, generic split, and generic join in order to abstract from the details and let the query processor figure out the match to that node in the inspected process model.

Figure 4.2 shows a sample BPMN-Q query along with a match to a process model, highlighted in gray. The query represents a path edge which connects two nodes, *A* and *D*, and returns the set of nodes that could exist in between these two nodes in the matching process model. BPMN-Q is also capable of expressing complex queries on the structure of process models. For example, nodes and path edges can be assigned with variable names or variable names which start with the symbol "?." These names can be used in the *exclude* property of paths to help describe nontrivial queries.

In practice, a query with a path from activity *Receive purchase request* to activity *Archive request* would be created by a business designer to look up situations of handling purchase requests. For the basic query processor of BPMN-Q, it looks for process models having activity labels strictly matching those in the query. Thus, process models having activities of the form *Get purchase order*, *Process purchase request*, etc., will not be inspected by the query processor, though they are *semantically* relevant to the query. In principle, the search with strict matching of activity labels would be sufficient in organizations with a high maturity level in composing processes. In this situation, process designers are strictly aligned with a rather fixed set of labels to describe their process steps. Unfortunately, this level of maturity is not the general case. Thus, semantic Web approaches were adopted and applied to the labeling of activities in order to either compose or discover processes [140]. However, this comes with the cost of manually analyzing process models and creating an ontology and assigning appropriate labels to activities before reasoning about them. To overcome the limitation of strictly matching activity labels of queries and processes, BPMN-Q employed information retrieval approaches to automate the discovery of *semantically* similar activities while avoiding the cost of manual labeling. The basic query processor was extended by a semantic expansion layer [37]. In that layer, information retrieval (IR) techniques are employed to analyze labels of activities in process models from a semantical point of view. The semantic expansion is lightweight as it derives the semantic similarity without using any prerequisite annotation of activities done by a human. Rather, a vector space model [411] with knowledge of WordNet[2] ontology is used on the words in the labels of activities to derive the similarity. With this expansion, models as discussed above are now relevant to the query. Of course, such an expansion adds to the complexity of the query processing. To control this complexity, the user is asked to determine a search *threshold* that controls the search depth and thus the time taken to process queries. Figure 4.4 represents examples of semantically similar queries that could be generated as an expansion for the query illustrated in Fig. 4.3. Although BPMN-Q is mainly focusing on applying structural-based similarity matching techniques, it should be noted that all string-based textual matching techniques on the task/activity level which have been introduced in Chap. 3 can be orthogonally aligned and integrated with BPMN-Q querying process.

Fig. 4.3 Query to look for loan handling processes

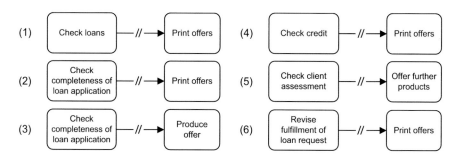

Fig. 4.4 Queries semantically similar to the one in Fig. 4.3

---

[2]http://wordnet.princeton.edu/

4.1.2 BP-QL

The BP-QL query language is based on an intuitive model that represents an abstraction of the BPEL specification [161], along with a graphical user interface that allows for simple formulation of queries over this model. In principle, it follows the same design principles for the graphical editors of the specification of BPEL processes. In addition, it hides from the users the tedious BPEL XML details and allows for more natural query formulation. In particular, the core of the BP-QL language is based on a set of patterns that allow users to describe the pattern of activities/data flow that are of interest. These patterns are similar to the tree and graph patterns offered by the existing query languages for XML (XPath[3]) and graph-shaped data [121, 371] but include two main features which are designed to address the needs for querying business process models:

- BP-QL supports navigation along two main axes:

  1. The standard path-based axis, which allows to navigate through, and query, paths in process graphs.
  2. A novel *zoom-in* axis, which allows to navigate (transitively) inside process components and query them at any depth of nesting.

- Paths are considered first-class objects in BP-QL and can be retrieved, and represented compactly, even when involving activities performed on distinct peers.

For querying business process models, BP-QL offers a set of querying patterns which play a similar role analogous to those patterns which are used for querying XML trees by tree pattern queries. In particular, BP-QL describes the pattern of activity/data flow that is of interest to the user and allows navigation along two axes: *path* based and *zoom-in* based. Following the use of / and // in XPath for denoting single- and multiple-step navigation, BP-QL patterns use edges with single and double heads to denote single- and multiple-edge paths, respectively. Similarly, to allow a user to query about flows that are nested at any depth in the zoom-in hierarchy, compound activity nodes may have doubly bounded boxes, to denote an unbounded zoom-in into the activities' internal specifications. The nodes and edges of BP patterns can be associated with variables, and these can be used in selection conditions on their attributes and data and for joins. BP-QL also supports negation operations which are denoted by dashed nodes and edges.

The BP-QL implementation is relying on using Active XML[4] (AXML for short) documents and compiling the input BP-QL queries into XQuery[5] queries over these documents. In AXML documents, some of the XML data is explicitly specified,

---

[3]http://www.w3.org/TR/xpath

[4]http://activexml.net/

[5]http://www.w3.org/TR/xquery/

while other parts are presented intentionally, by means of receiving the results of Web service calls. Therefore, when a query is evaluated over such documents, the calls relevant to the query execution are dynamically invoked. In particular, when a query is evaluated, its patterns are matched against the repository of business process models where the query nodes and edges are assigned activity/data/property nodes and execution/data flow paths, respectively. In particular, the semantics of a query q on repository S is defined as a function from the nodes and edges of q to nodes, edges, and paths of S, which satisfies the obvious constraints: nodes are mapped to nodes of the same type, and single-/double-head edges are mapped to edges/paths between the corresponding end points. When a compound query node is doubly bounded, nodes and edges in it may be mapped to nodes and paths in a process obtained by any number of zoom-ins into the activity's specification. For nodes and edges that are associated with variables, the query constraints on these variables must be satisfied as well.

4.1.3 BeehiveZ

BeehiveZ [242, 244] has been introduced as an approach for querying repositories of business process models where all the models in the repository are represented as or transformed to Petri nets and stored as text in a relational database management system (e.g., MySQL). BeehiveZ provides different indexes to speed up the query processing where all these indexes are inverted indexes which set up the mapping between the indexed items and the models. The indexed items are different according to different types of queries. These indexes can be used as filters to discard many models that are impossible to be resulting models, so that the query efficiency can be improved. In particular, BeehiveZ supports the following types of queries:

- *Exact query based on structure.* In general, subgraph isomorphism algorithm is NP-complete, and it is quite expensive to scan all the models in the repository sequentially for performing subgraph matching check between the input query and each stored model in the repository. Therefore, BeehiveZ uses path indexes as a filter to discard many models that are impossible to be the resulting models and obtains a set of candidate models whose size is always much smaller than the size of the repository and thus the query efficiency can be improved greatly.
- *Similarity query based on structure.* To find business process models which are sufficiently similarly matching to the input query, BeehiveZ uses the maximum common edge subgraph (MCES)-based similarity technique [41]. In general, two tasks which are having the label similarity greater than a specific threshold are regarded as identical (this threshold can be configured by users during query time). To process similarity queries efficiently, a label index is used to set up the mapping between words and labels where the investigated word appears. To retrieve synonyms quickly, BeehiveZ mapped *WordNet* synonyms into memory,

and during the query processing, the input query gets to expand queries with the synonymous labels existing in the repository with the help of the label index.

- *Exact query based on behavior.* BeehiveZ provides a language for users to describe the behavioral requirements of their queries and uses an index based on the ordering relations between tasks to efficiently improve the evaluation of these queries [241, 243].
- *Similarity query based on behavior.* In these types of queries, BeehiveZ measures the similarity between business process models based on their task adjacency relations using an index which is based on task adjacency relations that is used to facilitate the efficient query processing.

4.1.4 APQL

APQL [439] is *A Process* model *Query Language* which is designed to be independent of the actual used process modeling language (e.g., BPMN, EPCs). APQL is mainly focusing on the semantic relationships between tasks in process models by relying on a number of basic temporal relationships between tasks which can be composed to obtain complex relationships between them. These predicates allow the user to express queries that can discriminate over single process instances or task instances. In particular, APQL defines a set of basic predicates that capture the occurrences of tasks as well as the semantic relationships between tasks. Examples of these predicates are:

- *posoccur(t; r)*: some execution of process model (*r*) exists where at least one instance of task of type (*t*) occurs.
- *alwoccur(t; r)*: in every execution of process model (*r*), at least one instance of task of type (*t*) occurs.
- *exclusive(t1; t2; r)*: in every execution of process model (*r*), it is never possible that an instance of task of type (*t1*) and an instance of task of type (*t2*) both occur.
- *concur(t1; t2; r)*: an instance of task of type (*t1*) and an instance of task of type (*t2*) are not causally related, and in every execution of process model (*r*), if an instance of *t1* occurs, then an instance of *t2* occurs and vice versa.
- *succany(t1; t2; i)*: in process execution *i*, at least one instance of task of type (*t1*) occurs and is eventually succeeded by an instance of task of type (*t2*).
- *succevery(t1; t2; i)*: in process execution *i*, at least one instance of task of type (*t1*) occurs, and every instance of *t1* is eventually succeeded by an instance of task of type (*t2*).
- *predany(t1; t2; i)*: in process execution *i*, at least one instance of task of type (*t1*) occurs and is eventually preceded by an instance of task of type (*t2*).
- *predevery(t1; t2; i)*: in process execution *i*, at least one instance of task of type (*t1*) occurs, and every instance of *t1* is eventually preceded by an instance of task of type (*t2*).

While APQL has been proposed as an expressive language, there is no query execution engine for the language yet. In practice, when it comes to develop a technique for evaluating APQL queries, one main challenge is the ability of the query evaluation technique to determine the semantic relationships between task relationships in a feasible manner without suffering from the well-known state-space explosion problem.

4.2 Querying Business Process Execution

Large organizations may run hundreds or even thousands of business processes on a daily basis. In principle, querying running instances of business processes represents an important tool in the hand of the admin of a business process enactment engine to monitor the status of running processes and trace the progress of execution. In principle, a wide spectrum of use cases advocate the need for process monitoring. For example, achieving process execution flexibility is possible via monitoring process instances and responding to exceptional situations. In addition, processes have to be monitored for obeying regulations and responding immediately to violations. In between, the need to ensure satisfaction of service-level agreement and collecting information about the performance of processes and their loads are other interesting use cases. To illustrate, monitoring of process executions may allow the process administrator to be notified whenever an auctioneer cancels bids too often, or when buyers attempt to confirm bids without first giving their credit details, so that he/she can block their actions. Similarly, the process administrator can be notified whenever the average response time of the database in a given service passes a certain threshold, allowing him/her to fix the problem or switch to a backup database.

In practice, in many situations, different activities have to be executed under different conditions through a process instance lifetime. These conditions relate to process instance specific data, context data, and resources available for executions. Several approaches have been proposed in literature for enabling querying business process instances and attempting to provide instant knowledge about the state of the active instances of each process models such as: *BP-Mon* [52, 53] and *FPSPARQL* [56, 59]. In the following sections we give an overview and discuss the different capabilities and design decisions for each of these approaches.

4.2.1 BP-Mon

The BP-Mon (BP monitoring) query language [52, 53] presents a high-level graphical query language that allows for simple description of the execution patterns (EX-patterns) of BPEL-based business processes to be monitored. In particular,

BP-Mon queries consist of two main ingredients:

1. *EX-patterns* that are matched to execution traces
2. *Reports* which are generated based on the matches of the EX-patterns

In principle, EX-patterns are used by BP-Mon for monitoring process instances. EX-patterns extend string regular expressions to (nested) DAGs. The constructs of the query EX-patterns look much like the BPEL specifications (e.g., while, switch, etc.) in addition to two querying constructs: *or* describing alternative patterns and *rep* describing repetitions. Similar to BP-QL [51], the patterns also allow to navigate in the activities' flow along two axes: path based and zoom-in based. In addition, the activities and edges of EX-patterns can be associated with variables, which can be used in selection conditions on the values of the associated attributes/data variables and in reports. To issue a *report*, a user notification facility can be connected to a reporting point in the pattern (an atomic or a compound activity). In principle, a BP-Mon query may include several such reporting icons/points. Two reporting modes are available:

1. *Local*: where an individual report is issued for each process instance.
2. *Global*: this mode considers all the business process instances.

For each report, the user can specify when should it be issued (e.g., at the first time that the reporting point is reached, at periodic time interval, or when certain conditions are satisfied) and what should be the structure of the output (in XML format) or the actions triggered at this point.

On the low level, on one hand, the execution trace of a BPEL process instance can be viewed as a DAG where each activity is represented by a pair of time-stamped nodes, corresponding to its activation and completion. For a compound activity, the DAG that represents its internal flow appears (time-wise) between its activity activation and completion nodes and is connected to them by zoom-in edges. On the other hand, BP-Mon queries are modeled by EX-patterns that generalize execution traces (EX-traces) similarly to the way tree patterns generalize XML trees. In principle, EX-patterns can be considered as EX-traces without time stamps where node labels are either specified or left open using a special *ANY* symbol and where two additional new label symbols can be used: *or* and *rep*. Edges in a graph are either regular edges, interpreted over edges, or transitive, interpreted over paths. Similarly, activity pairs may be regular or transitive, for searching only in their direct internal trace or zoom-in transitively inside it.

To ease the job of the users, BP-Mon queries are written via a visual editor. The query translator compiles a query on p to a BPEL process which is then deployed onto the BPEL server where the instances of p are executed. Several queries monitoring the same or different processes may be deployed on a server. Using an XML stream processing system, a report is generated when a successful matching for the query pattern associated with a report node generates an XML entry recording the embedding and the *Report* command is applied to this stream of matches.

4.2.2 BP-SPARQL

Beheshti et al. [56, 61] have presented a business process event query language called *BP-SPARQL*, which is a Folder-Path-enabled extension of SPARQL [160]. BP-SPARQL is based on a data model which represents a process execution log as a graph of typed nodes and edges. This data model is based on three main concepts:

1. *Entity*: which is represented as a data object that exists separately and has a unique identity.
2. *Folder node*: which contains entity collections. In particular, each folder node represents a grouping for a collection of related entities based on a certain pattern or query.
3. *Path node*: which refers to one or more paths in the graph where a path represents the transitive relationship between two entities.

Based on this data model, BP-SPARQL supports two levels of queries:

1. *Entity-level queries*: which rely on the standard SPARQL constructs to query entities of the business process logs
2. *Aggregation-level queries*: which rely on the extended constructs of BP-SPARQL to construct and query folder and path nodes in addition to grouping related nodes and edges of the executed business processes

The BP-SPARQL querying mechanism uses a relational database system for storing the events of the executed business processes and relies on SPARQL-to-SQL translation and query optimization algorithms for optimizing the performance of the query engine [109, 110, 409]. The query engine is also equipped with a graphical query tool that supports the users for expressing their queries.

4.3 Business Process Compliance

Today's enterprises demand a high degree of compliance of business processes to meet diverse regulations and legislations. Several industrial studies have shown that compliance management is a daunting task, and organizations are still struggling and spending billions of dollars annually to ensure and prove their compliance. In particular, the global regulatory environment has grown in complexity and scope since the financial crisis in 2008. This is causing significant problems for organizations in almost all industrial sectors, as the complexity of hard and soft regulations has been little understood or appreciated. For example, banking regulations such as anti-money laundering directives are generally complex and far-reaching, with a raft of major banks found not to be in compliance in 2012. Standard

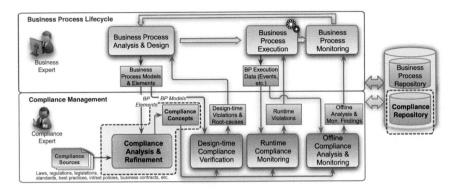

Fig. 4.5 Framework architecture for business process compliance management [446]

Chartered Bank, London, for example, was fined a total of $459 million in 2012.[6] Worse still HSBC Holdings Plc. had paid a record of $1.92 billion. These incidents were preceded by scandals and business failures such as Enron and WorldCom back in 2001. Subsequently, much attention has been paid to compliance management from both the academic and the industrial communities.

Several techniques for business process compliance management have been proposed to check and enforce compliance of business processes. Türetken et al. [446] have presented a general approach and architecture for business process compliance framework. The proposed approach relies on two logical repositories: the *business process repository* and the *compliance repository*. Figure 4.5 illustrates the proposed architecture which may start from the business process life cycle (the upper part of the architecture) or with compliance management practices (the lower part of the architecture) which afterward align and run together exchanging inputs and outputs. In practice, business process compliance mechanisms involve the utilization of business process querying and matching techniques. However, their main goals go further than that. In particular, business process life cycle starts with the analysis and design of the processes. On the other side, the compliance management practices commence with the compliance analysis and refinement, which involves, first, the analysis of the sources of compliance requirements, such as laws, regulations, standards, policies, etc., that state the norms mandating or impacting the way the business processes are executed, and, second, the transformation of these abstract norms into a set of concrete concepts relevant to compliance management. In practice, the architecture covers three main compliance assurance activities each having a corresponding business process life cycle stage. *Design-time compliance verification* involves the static verification of business process models against formal compliance rules. *Design-time compliance violations* and

---

[6]http://www.accuity.com/industry-updates/free-resources/trends-in-aml-compliance-infographic/

possible *root causes* provide key input for business process analysis and design activities to ensure that business process models progressing to execution are compliant by design. Compliance checks at design time are critical; however, it is not always feasible to enforce compliance with all constraints imposed on a process model at design time. *Runtime compliance monitoring* (i.e., verifying compliance dynamically during business process execution) and *offline compliance analysis and monitoring* (i.e., verifying compliance after business process execution) are vital for a holistic view ensuring compliance throughout the remaining phases of the business process life span. During runtime compliance monitoring, the execution of the business process model instances is observed against runtime compliance rules. Therefore, apparently, achieving effective process compliance management involves a lot more than querying process models or their instances.

Formal-Based Compliance Schemes In general, runtime monitoring requires business process models to be reduced to some abstract representation, which are built up by collecting runtime information (e.g., exchanged message sequences, performed activities). On the other hand, runtime monitoring also requires compliance requirements to be structurally/formally represented using a formal/structural language, e.g., linear temporal logic (LTL), computational tree logic (CTL), and event-condition-action (ECA) rules. Formal monitoring approaches [44, 209, 313, 340] have founded their compliance requirement techniques on formal/mathematical languages. For example, Barbon et al. [44] presented an approach that provides the ability to monitor both the behaviors of single instances of BPEL processes in addition to monitoring the behaviors of a class of instances. In particular, the monitors can check temporal, Boolean, time-related, and statistic properties. The approach relies on its own defined language which is expressive enough to formally represent the specifications of all these kinds of monitors. The defined monitors are then automatically compiled to generate both instance and class monitors from their specifications. Hallé and Villemaire [209] have introduced *LTL-FO+* for expressing data-aware properties of execution traces. In particular, LTL-FO+ represents an augmentation of a LTL with full first-order quantification capabilities over the data inside a trace of XML messages. Mahbub and Spanoudakis [313] used *event calculus* (EC) as the formal basis of monitored constraints against BPEL models. In principle, EC is a logic language based on first-order predicate calculus that can be used to represent and reason about the behavior of dynamic systems. EC is an expressive language; however, it is excessively difficult to be used due to its formalized specification. Monitoring is implemented as integrity-checking techniques on *completed* executions. *MOBUCON* [340] is a monitoring framework which has also used event calculus (EC) to specify the constraints to be monitored. The framework has been embedded inside the operational decision support infrastructure of *ProM*.[7] In order to cope with the complexity of EC, MOBUCON has utilized the *declare* language [377] as a graphical intermediate

---

[7]http://www.promtools.org/

representation. Logic programming reasoning is then used to dynamically reason about partial, evolving execution traces.

XML Querying-Based Compliance Schemes XML querying techniques have been one of the main directions for implementing business process compliance techniques. In general, one of the advantages of using XML querying techniques is their standardization on dealing with XML traces that can be generated from any XML-based business process execution environment (e.g., BPEL). For example, *BPath* [417] is an XPath-based language for both expressing and checking temporal and hybrid logical properties over execution traces of business processes at runtime. Hallé and Villemaire [210] have presented another approach that performs trace validation of LTL formulae by relying on readily available XML technologies. In particular, they provide a translation mechanism between LTL and a subset of the XML query language, *XQuery*,[8] and show that an efficient validation of LTL formulae can be achieved through the evaluation of XQuery expressions. Furthermore, they show that the translation mechanism can be extended to support LTL-FO+ [209].

Complex Event-Processing-Based Compliance Schemes Several approaches have employed the *complex event-processing* (CEP) technology for implementing business process compliance mechanism [350, 442, 467]. In general, CEP represents a set of tools and techniques for analyzing and controlling a complex series of interrelated events. In particular, events are observed as they occur and are correlated in order to discover and respond to certain event patterns. Therefore, in practice, business process compliance techniques are usually designed on event pattern languages (EPLs) that enable expressing and capturing the relevant compliance constraints. For example, Mulo et al. [350] have presented an approach where model-driven engineering techniques have been adopted by introducing a high-level DSL for representing the abstract specification of compliance constraints. In particular, the proposed domain-specific language is used for representing the specification of compliance requirements. Code-generation templates are used to generate the compliance monitoring code, based on the DSL-based specifications, which is then leveraged by CEP engine to monitor the specified compliance rules. Weidlich et al. [467] have presented another CEP-based approach that leverages the concept of a behavioral profile to generate queries from a process model and supports the presentation of the query results in such a way that root causes of a rule violation are directly visible to the process analyst.

Thullner et al. [442] presented an approach that models business processes as event flows where compliance requirements are structurally represented in a rule model which is proposed by the authors. In this approach, compliance checkpoints are added to the event flow with the aim of signifying the aspects which may be relevant to monitor such as the relative time frame between two events. The defined monitoring rules are used to detect compliance violations using the underlying

---

[8]http://www.w3.org/TR/xquery/

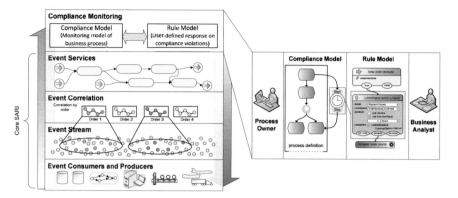

Fig. 4.6 Business process compliance monitoring with SARI [304]

CEP engine and automatically trigger corrective actions. In particular, the proposed compliance framework is an extension of the event-based system SARI (Sense and Response Infrastructure) [349], which is a CEP system that allows users to compose complex pattern-detection logic by visually modeling event-triggered decision graphs. Figure 4.6 illustrates an overview of SARI's high-level architecture and the extensions for compliance monitoring. According to this architecture, events are received from and propagated to various event producers and consumers. Incoming events can be correlated based on event attributes and therefore it is possible to identify which events belong to the same instance of a business process. The business logic of a SARI application is implemented in several event services, where each service encapsulates a part of the overall business logic. Incoming events are passed to event services that process them according to their business logic. Event services are then orchestrated to build the complete business logic of an SARI application.

Graph-Based Compliance Schemes The SeaFlows toolset [304] provides a user-friendly environment for modeling compliance rules using a graph-based formalism. In addition, SeaFlows provides two compliance checking components: the *structural compliance checker* that derives structural criteria from compliance rules and applies them to detect incompliance and the *data-aware compliance checker* that addresses the state explosion problem that can occur when the data dimension is explored during compliance checking. The checker performs context-sensitive automatic abstraction to derive an abstract process model which is more compact with regard to the data dimension enabling more efficient compliance checking. Figure 4.7 illustrates the architecture of the SeaFlows toolset. In particular, the graphical compliance rule editor allows to model compliance rules over process artifacts in the form of compliance rule graphs which provide visual representation, enable the representation of common compliance rule patterns, and are particularly suited for compliance monitoring. The SeaFlows compliance checkers enable the process designer to already verify process models during process design. Thus,

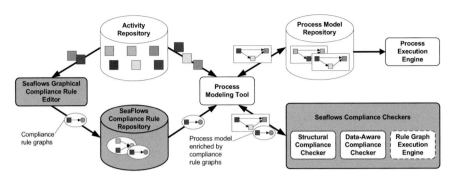

Fig. 4.7 The architecture of the SeaFlows toolset [304]

meaningful compliance reports help the process designer to identify incompliant process behavior. Based on them, the process designer may further modify the process model until incompliance is resolved. The compliance checker of the SeaFlows toolset uses model-based diagnosis theory that permits to discover the event responsible of a malfunction comparing the model that describes the system (expected behavior) with the observational model (observed behavior).

Chapter 5
Business Process Data Analysis

The problem of understanding the behavior of information systems as well as the processes and services they support has become a priority in medium and large enterprises. This is demonstrated by the proliferation of tools for the analysis of process executions, system interactions, and system dependencies and by recent research work in process data warehousing and process discovery. Indeed, the adoption of business process intelligence (BPI) techniques for process improvement is the primary concern for medium and large companies. In this context, identifying business needs and determining solutions to business problems require the analysis of business process data. Analysis of business data will help in discovering useful information, suggesting conclusions, and supporting decision-making for enterprises and enable a process analyst to answer questions such as: *where are the bottlenecks in the purchasing process? what is the actual process typically followed for invoice payment? what is the status of purchase order number 325, who processed it and how? and how to find all the information related to a specific purchase of a customer (order, invoice, payment, shipping, etc.) in the enterprise?*

In this chapter, we give an overview of different aspects of business data analysis techniques and approaches from process/dataspaces to data provenance and data-based querying techniques. We start with an overview of *warehousing process data* followed by introducing *data services* and *DataSpaces* which facilitate organizing and analyzing process-related data. Next we discuss the importance of supporting big data analytics over process execution data. Afterward we define a holistic view of the process executions over various information systems and services (i.e., process space) followed by a brief overview of process mining to highlight the interpretation of the information in the enterprise in the context of process mining. Finally, we focus on process artifacts and introduce crosscutting aspects in processes data and discuss how process analytics can benefit from crosscutting aspects such as provenance, e.g., to analyze the evolution of business artifacts.

© Springer International Publishing Switzerland 2016   107
S.-M.-R. Beheshti et al., *Process Analytics*, DOI 10.1007/978-3-319-25037-3_5

5.1   Warehousing Business Process Data

Improving business processes is critical to any corporation. Process improvement
requires analysis as its first basic step. Before analyzing the process data, there is
a need to capture and organize the process data. This is important as executions of
process steps, in modern enterprises, leave temporary/permanent traces in various
systems and organizations. In order to analyze process data, it is possible to collect
the data into a data warehouse, using extract, transform, and load (ETL) tools, and
then leverage an OLAP tool to slice and dice data along different dimensions [102].

In this context, the process data warehousing presents interesting challenges
[102]:

(a) Outsourcing: developing an ad hoc and process-specific solution for warehous-
    ing and reporting on process data are not a sustainable model.
(b) Process data abstraction: the typical process executed in the IT system is very
    detailed and consists of dozens of steps, including manual operations (e.g.,
    scanning invoices), database transactions, and application invocations.
(c) Data evolution: the business process automation/analysis application is code-
    veloped, which means that, during development, changes to the data sources
    and even to the reporting requirements are fairly frequent.

Considering the abovementioned challenges, and in the domain of business
processes, it is important to devise a method for minimizing the impact of changes
and be able to quickly modify and retest the ETL (extract, transform, and load)
procedures, the warehouse model, and the reports. To address these challenges,
Casati et al. [102] proposed a conceptual model for process data warehousing. In
particular, they provided a configurable warehouse model that can satisfy complex
reporting needs for virtually any process, also taking into account performance
constraints. The model addresses key recurring problems such as the trade-off
between the need to model heterogeneity (each process is different) and that of
defining a uniform representation for all processes (to support reusability and cross
process analysis).

To support warehousing for business process data, there is a need to provide users
with a way to model the abstraction. This will help in understanding the high-level
processes and also it will describe how its progression maps to underlying IT events.
Moreover, there is a need for an ETL mechanism that, based on the abstract process
definition and the events occurring on the different systems, loads the warehouse
with abstracted process execution data. To address these requirements, modeling
abstract processes should involve: (a) describing the process flow, (b) specifying
how the abstracted business data for each process is populated and maintained,
(c) associating the start and completion of each step with changes to the abstract
business data, and (d) associating steps to human or automated resources.

To populate the process data warehouse, it is necessary to first extract the
data from the different event log databases into the landing tables of the staging
area. In the process domain, data services play an important role in extracting

process-related data. Following we explain data services and we discuss how they provide an emulation environment that supports testing and prototyping of events and data: once a process is started, each step binds to a data service and will be assigned for execution to a data generation Web service.

5.1.1 Data Services

In the enterprise world, data services play an important role in SOA architectures [98, 99, 202, 399]. For example, when an enterprise wishes to controllably share data (e.g., structured data such as relational tables, semi-structured information such as XML documents, and unstructured information such as commercial data from online business sources) with its business partners, via the Internet, it can use data services to provide mechanisms to find out which data can be accessed, what are the semantics of the data, and how the data can be integrated from multiple enterprises. In particular, data services are "software components that address these issues by providing rich metadata, expressive languages, and APIs for service consumers to send queries and receive data from service providers" [99].

A Web service, i.e., a method of communication between two electronic devices over the Web [29], can be specialized, as a data service, to encapsulate a wide range of data-centric operations, where these operations need to offer a semantically richer view of their underlying data in order to use or integrate entities returned by different data services [99]. Microsoft's WCF Data Services framework,[1] which enables the creation and consumption of OData services for the Web, and Oracle's ODSI,[2] which provides a wide array of data services designed to improve data access from disparate data sources for a wide range of clients, are two of a number of commercial frameworks that can be used to achieve this goal.

In this context, SOA applications will often need to invoke a service to obtain data, operate locally on that data, and then notify the service of changes that the application wishes to make to the data. Consequently, standards activity is needed in the context of data services. For example, the Open SOA Collaboration's Service Data Objects (SDO) specification [399] addresses these needs by defining client-side programming models, e.g., for operating on data retrieved from a data service and for XML serializing objects, and their changes for transmission back to a data service [98]. In particular, the use of data is bound to various rules imposed by data owners, and the (data) consumers should be able to find and select relevant data services as well as utilize the data "as a service."

Data as a service, or DaaS, is based on the concept that the data can be provided on demand to the user regardless of geographic or organizational separation of provider and consumer [445]. In particular, data services are created to integrate

---

[1] http://msdn.microsoft.com/en-us/data/bb931106

[2] http://docs.oracle.com/cd/E13162_01/odsi/docs10gr3/

as well as to service enable a collection of data sources. These services can be used in mash-ups, i.e., Web applications that are developed starting from contents and services available online, to use and combine data from two or more sources to create new services. In particular, data services will be integral for designing, building, and maintaining SOA applications [98]. For example, Oracle's ODSI supports the creation and publishing of collections of interrelated data services, similar to *dataspaces*.

Data services can be leveraged to reduce the effort required to set up a data integration system and to improve the system in "pay-as-you-go" fashion as it is used. In this context, data integration approaches require semantic integration before any services can be provided. It is important as process data is scattered across several systems and data sources and there is no single schema to which all the process-related data conforms. To address this challenge, *Dataspaces* are proposed to overcome some of the problems encountered in data integration systems and to promote awareness of the data and address concerns for ensuring the long-term availability of data in repositories.

5.1.2 DataSpaces

Dataspaces are an abstraction in data management that aim to manage a large number of diverse interrelated data sources in enterprises in a convenient, integrated, and principled fashion. Dataspaces are different from data integration approaches in a way that they provide base functionality over all data sources, regardless of how integrated they are. For example, a dataspace can provide keyword search over its data sources, and then more sophisticated operations (e.g., mining and monitoring certain sources) can be applied to queried sources in an incremental, pay-as-you-go fashion [206]. These approaches do not consider the business process aspects per se; however, they can be leveraged for organizing and managing ad hoc process data.

DataSpace Support Platforms (DSSPs) have been introduced as a key agenda for the data management field and to provide data integration and querying capabilities on (semi-)structured data sources in an enterprise [206, 413]. For example, SEMEX [94] and Haystack [252] systems extract personal information from desktop data sources into a repository and represent that information as a graph structure where nodes denote personal data objects and edges denote relationships among them.

The design and development of DSSPs have been proposed in [168]. In particular, a DSSP [168, 206, 413]:

- Helps to identify sources in a dataspace and interrelated identified resources. A DSSP is required to support all the data in the dataspace rather than leaving some out, as with DBMSs.
- Offers basic searching, querying, updating, and administering mechanisms over resources in a dataspace, including the ability to introspect about the contents.

However, unlike a DBMS, a DSSP is not in full control of its data, as the same data may also be accessible and modifiable through an interface native to the system hosting the data.

- Does not require full semantic integration of the sources in order to provide useful services: there is not a single schema to which all the data conforms and the data resides in a multitude of host systems.
- Offers a suite of interrelated (data integration and querying) services in order to enable developers focusing on the specific challenges of their applications. Queries to a DSSP may offer varying levels of service, as sometimes individual data sources are unavailable and best-effort or approximate answers can be produced at the time of the query.
- Provides mechanisms for enforcing constraints and some limited notions of consistency and recovery, i.e., to create tighter integration of data in the space as necessary.

Motivating applications for DSSPs include scenarios in which related data are scattered across several systems and data sources, e.g., personal information management systems [152], which are used to acquire, organize, maintain, retrieve, and use information items (e.g., desktop documents, Web pages, and email messages) accessed during a person's lifetime, and scientific data management systems [191], which are used for record management for most types of analytical data and documentation which ensures long-term data preservation, accessibility, and retrieval during a scientific process.

In order to search and query dataspaces, a new formal model of queries and answers should be specified. This is challenging as answers will come from multiple sources and will be in different data models and schemas. Moreover, unlike traditional querying/answering systems, a DSSP can also return sources, i.e., pointers, to places where additional answers can be found. Some works [204, 322] presented semantic mapping techniques to reformulate queries from one schema to another in data integration systems. Another line of related work [77, 201] focused on ranking answers in the context of keyword queries to handle the heterogeneity of resources. Some other works, e.g., in [288], focused on finding relevant information sources in large collections of formally described sources.

In dataspaces, a significant challenge is to answer historical queries which applied to heterogenous data. A line of research proposed techniques for modeling and analyzing provenance [112] (also known as lineage and pedigree), uncertainty, and inconsistency of the heterogenous data in dataspaces [237, 472]. Many provenance models [112, 170, 342, 421] have been presented, motivated by notions such as influence, dependence, and causality in such systems. Moreover, the relationship between uncertainty and provenance is discussed in [472].

Dataspaces are large collections of heterogeneous and partially unstructured data, and therefore, indexing support for queries that combine keywords and the structure of the data can be challenging. For example, in [145], authors proposed an indexing technique for dataspaces to capture both text values and structural information using an extended inverted list. Their proposed framework extends inverted lists that capture attribute information and associations between data items, i.e., to support robust indexing of loosely coupled collections of data in the presence of varying degrees of heterogeneity in schema and data. Another indexing system [134] is designed to provide entity search capabilities over datasets as large as the entire "Web of data." Their approach supports full-text search, semi-structural queries, and top-k query results while exhibiting a concise index and efficient incremental updates. Challenges in implementing a scalable and high-performance system for searching semi-structured data objects over a large heterogeneous and decentralized infrastructure have been discussed in [133], where an indexing methodology for semi-structured data have been introduced.

Recently, a new class of data services is designed for providing data management in the cloud [459]: the cloud is quickly becoming a new universal platform for data storage and management. In practice, data warehousing, partitioning, and replication are well-known strategies to achieve the availability, scalability, and performance improvement goals in the distributed data management world. Moreover, a database as a service is proposed as an emerging paradigm for data management in which a third-party service provider hosts a database as a service [202]. Data services can be employed on top of such cloud-based storage systems to address challenges such as availability, scalability, elasticity, load balancing, fault tolerance, and heterogenous environments in data services. For example, Amazon Simple Storage Service (S3) is an online public storage Web service offered by Amazon Web Services.[3]

A growing number of organizations have begun turning to various types of non-relational, *NoSQL* (not only SQL) databases such as Google BigTable [106], Yahoo PNUTS [123], and Amazon Dynamo [132]. NoSQL is a broad class of low-cost and high-performance database management systems and proposed to address RDBMSs' shortcomings: ever-increasing needs for scalability and new advances in Web technology, which requires facilitating the implementation of applications as a distributed and scalable services, have created new challenges for RDBMSs [426, 459]. Such databases are designed to be very scalable and reliable, and they consist of thousands of servers geographically distributed all over the world. Major research challenges for providing heterogenous data management, e.g., using data services, need [98, 99, 202]:

(a) Dynamically reconfigurable runtime architectures: distributed service components and resources should be leveraged to create an optimal architectural configuration to both the particular user requirements and the application characteristics.
(b) End-to-end security solutions: a full system approach to test end-to-end security solutions at both the network and application level is required.

---

[3]http://aws.amazon.com/

(c) The infrastructure support for data and process integration: uniform consistent access to all heterogenous data should be provided, i.e., irrespective of the data format, source, or location.
(d) The analytic support for the discovery and communication of meaningful patterns in (process execution) data, e.g., *business analytics*.

The field of business analytics has improved significantly over the past few years, giving business users better insights, particularly from operational data managed by dataspaces. For example, banks that developed an analytic application for budgeting and forecasting targeted at the financial service industry determined that its online analytical processing (OLAP), or OLAP, can provide the capability for complex calculations, trend analysis, and sophisticated data modeling. In particular, OLAP can be used to reduce the time needed for analyzing the process data by providing powerful user interfaces that let the analyst explore the process-related data along previously defined analysis dimensions.

5.2 Supporting Big Data Analytics over Process Execution Data

In modern enterprises, businesses accumulate massive amounts of data from a variety of sources. In order to understand businesses, one needs to perform considerable analytics over large hybrid collections of heterogeneous and partially unstructured process-related execution data. These data increasingly come to show all typical properties of the *big data*: wide physical distribution, diversity of formats, nonstandard data models, and independently managed and heterogeneous semantics, and need to be represented as graphs, i.e., *big process graphs*. The discovery and communication of meaningful patterns in data (i.e., analytics) can help in understanding the big business data with an eye to predicting and improving business performance in the future.

In order to understand available data (events, business artifacts, data records in databases, etc.) in the context of process execution, we need to represent them, understand their relationships, and enable the analysis of those relationships from the process execution perspective. To achieve this, it is possible to represent process-related data as entities and any relationships among them (e.g., event relationships in process logs with artifacts) in entity-relationship graphs. In this context, business analytics can facilitate the analysis of process graphs in a detailed and intelligent way through describing the applications of analysis, data, and systematic reasoning [38, 93, 219, 265]. Consequently, an analyst can gather more complete insights using techniques such as modeling, summarizing, and filtering.

Applications of business analytics extend to nearly all managerial functions in organizations. For example, considering financial services, applying business analytics on customer dossiers and financial reports can specify the performance of the company over periods of time. As another example, consider the collaborative relationship between researchers, affiliated with various organizations, in the process of writing scientific papers, where it would be interesting to analyze the collaboration patterns [27, 28] (e.g., frequency of collaboration, degree of collaboration, mutual impact, and degree of contribution) among authors or analyze the reputation of a book, an author, or a publisher in a specific year. Such operations require supporting n-dimensional computations on process graphs, providing multiple views at different granularities, and analyzing set of dimensions coming from the entities and the relationship among them in process graphs.

In traditional databases (e.g., relational DBs), data warehouses and OLTP (online transaction processing) and OLAP (online analytical processing) technologies [15, 107] were conceived to support decision-making and multidimensional analysis within organizations. To achieve this, a plethora of OLAP algorithms and tools have been proposed for integrating data, extracting relevant knowledge, and fast analysis of shared business information from a multidimensional point of view. Moreover, several approaches have been presented to support the multidimensional design of a data warehouse. Cubes defined as a set of partitions are organized to provide a multidimensional and multilevel view, where partitions are considered as the unit of granularity. Dimensions are defined as perspectives used for looking at the data within constructed partitions. Furthermore, OLAP operations have been presented for describing computations on cells, i.e., data rows.

While existing analytics solutions, e.g., OLAP techniques and tools, do a great job in collecting data and providing answers on known questions, key business insights remain hidden in the interactions among objects and data: most objects and data in the process graphs are interconnected, forming complex, heterogeneous but often semi-structured networks. Traditional OLAP technologies were conceived to support multidimensional analysis; however, they cannot recognize patterns among process graph entities, and analyzing multidimensional graph data, from multiple perspectives and granularities, may become complex and cumbersome. Existing approaches [57, 111, 162, 214, 248, 262, 388], in online analytical processing on graphs, took the first step by supporting multidimensional and multilevel queries on graphs; however, much work needs to be done to make OLAP heterogeneous networks a reality [215]. The major challenges here are (a) how to extend decision support on multidimensional networks, e.g., process graphs, considering both data objects and the relationships among them, and (b) providing multiple views at different granularities is subjective: it depends on the perspective of OLAP analysts how to partition graphs and apply further operations on top of them.

Besides the need to extend decision support on multidimensional network in process data analysis scenarios, the other challenge is the need for scalable analysis techniques. Similar to scalable data processing platforms [483], such analysis and querying methods should offer automatic parallelization and distribution of large-scale computations, combined with techniques that achieve high performance on large clusters, e.g., cloud-based infrastructure, and be designed to meet the challenges of process data representation that should capture the relationships among data (mainly, represented as graphs). In particular, there is a need for new scalable and process-aware methods for querying, exploration, and analysis of process data in the enterprise because: (a) process data analysis methods should be capable of processing and querying a large amount of data effectively and efficiently and therefore have to be able to scale well with the infrastructure's scale and (b) the querying methods need to enable users to express their data analysis and querying needs using process-aware abstractions rather than other lower-level abstractions.

To address these challenges, P-OLAP [61] (process OLAP) is proposed to support scalable graph-based OLAP analytics over process execution data. The P-OLAP goal is to facilitate the analytics over big process graph through summarizing the process graph and providing multiple views at different granularities. P-OLAP benefits from BP-SPARQL [54] (business process SPARQL), a MapReduce-based graph processing engine, for supporting big data analytics over process execution data. The P-OLAP framework has been integrated into ProcessAtlas [54], a process data analytics platform, which introduces a scalable architecture for querying, exploration, and analysis of large process data.

5.2.1   Online Analytical Processing

There is an analytical orientation to the nature of the process data, and consequently process analytics can benefit from decision support systems and business intelligence tools. Consequently, process analytics can benefit from OLTP and OLAP to reduce the time needed for analyzing the process data. This is important as the large amount of process-related data generated every second needs to be analyzed in almost real time. OLTP proposed to facilitate and manage transaction-oriented applications, typically for data entry and retrieval transaction processing. OLAP proposed to support analysis and mining of long data horizons and to provide decision-makers with a platform from which to generate decision-making information. In this section we focus on OLAP environments and discuss its importance in process analytics.

OLAP applications typically access large (traditional) databases using heavy-weight read-intensive queries. OLAP encompasses *data decision support* (focusing on interactively analyzing multidimensional data from multiple perspectives) and *data mining* (focusing on computational complexity problems). There have been a lot of works, discussed in a recent survey [401] and a book [441], dealing with multidimensional modeling methodologies for OLAP systems. Multidimensional conceptual views allow OLAP analysts to easily understand and analyze data in terms of facts (the subjects of analysis) and dimensions showing the different points of view where a subject can be analyzed from. These lines of works propose OLAP data elements such as partitions, dimensions, and measures and their classification, e.g., classifying OLAP measures into distributive, algebraic, and holistic. They discuss that one fact and several dimensions to analyze it give rise to what is known as the data cube.

There are many works, e.g., [185, 213], dealing with the efficient computation of OLAP data cubes, including: (a) efficient methods for computing iceberg cubes[4] [213] with some popularly used measures, such as average; (b) efficiently computing multiple related skyline[5] results; (c) computing closed iceberg cubes more efficiently using aggregation-based approach; and (d) proposing an algebra that operates over data cubes, independently of the underlying data types and physical data representation [185].

Many other works, e.g., [23, 173, 297], deal with clustering and partitioning of large databases, including: (a) presenting a classification of OLAP queries to decide whether and how a query should be parallelized [23]; (b) proposing an efficient solution, called adaptive virtual partitioning (AVP), for parallel query processing in a database cluster [297]; (c) combining the physical and virtual partitioning to define table subsets in order to provide flexibility in intra-query parallelism; (d) analyzing independent data tuples that mathematically form a set, i.e., conventional spreadsheet data; (e) leveraging both spreadsheets and ad hoc OLAP tools to assess the effects of hypothetical scenarios [43]; and (f) clustering and classification of graphs by studying systematically the methods for mining information networks and classifying graphs into a certain number of categories by similarity [215]. All these works provide some kind of (network) summaries incorporating OLAP-style functionalities.

Other works [56, 61, 151, 171, 388] focused on mining and querying information networks, including:

(a) Proposing techniques for query processing and cube materialization on infor-mational networks [388].
(b) Defining constraints on nodes and edges simultaneously on the entire object of interest, not in an iterative one-node-at-a-time manner. Therefore, they do not support querying nodes at higher levels of abstraction [151, 171].
(c) Proposing summarization frameworks to facilitate the analysis of process data modeled as graphs [56].
(d) Facilitating the analytics over big process graph, P-OLAP [61], through summa-rizing the process graph and providing multiple views at different granularities.

5.2.2 Trend, What-If, and Advanced Analysis

Various methods and techniques have been proposed for analysis and interpretation of process data. The focus of these techniques is on the behavior of completed processes, evaluating currently running process instances, and predicting the behavior of process instances in the future. Some of these techniques [2, 6, 53, 339] are

---

[4]An iceberg cube contains only those cells of the data cube that meet an aggregate condition. It is called an iceberg cube because it contains only some of the cells of the full cube [76]

[5]Skyline [478] has been proposed as an important operator for multi-criteria decision-making, data mining and visualization, and user preference queries

purely syntax oriented, focusing on filtering, translating, interpreting, and modifying event logs given a particular question. Other methods [89, 100, 223, 324] focused on the semantics of process data and tried to propose techniques to understand the hidden relationships among process artifacts. In particular, existing works on business analytics focused more on exploration of new knowledge and investigative analysis using a broad range of analysis capabilities, including: trend analysis, what-if analysis, and advanced analysis.

The focus in trend analysis is to explore data and track business developments with capabilities for tracking patterns. For example, it is possible to track patterns in Web services, as services leave trails in so-called event logs and recent break-throughs in process mining research make it possible to discover, analyze, and improve business processes based on such logs [4]. Also it is possible to monitor the status of running processes and trace the progress of execution [53, 339] or enable semantic process mining in order to track business patterns [89, 100, 223, 324]. Some of these techniques, e.g., [89, 324], are implemented as plug-ins in the ProM framework tool.

In what-if analysis, scenarios with capabilities for reorganizing, reshaping, and recalculating data are of high interest. In this category, business process data can be used to forecast the future behavior of the organization through techniques such as scenario planning and simulation [108]. One research direction is to explore the relationships between what-if analysis and multidimensional modeling [269] and to analyze the natural coupling, which exists between data modeling, symbolic modeling, and what-if analysis phases of a decision support systems. Another line of research describes what-if analysis as a data-intensive simulation to inspect the behavior of a complex system under some given hypotheses [184]. Also it is possible to perform what-if analysis, by investigating the requirements for a process simulation environment.

Advanced analysis techniques provide techniques to uncover patterns in businesses and discover relationships among important elements in an organization's environment. Linking between entities across repositories has been the focus of a large number of works. For example, the idea of linked data[6] has recently attracted a lot of attention in information systems, where research directions that include: discovery of semantic links from data based on declarative specification of linkage requirements by a user [219], link semantically related entities across internal and external data sources using the power of external knowledge bases available on the Web [218], investigating the problem of event correlation for business processes [346] to detect correlation identifiers from arbitrary data sources in order to determine relationships between business data [403] and to discover inter-process relationships in a process repository [277].

Moreover, a new stream of work [91, 187, 318] has focused on weaving social technologies to business process management. They aim to consolidate

---

[6]Linked data is a method of publishing data on the Web based on principles that significantly enhance the adaptability and usability of data, either by humans or machines [84]

the opportunities for integrating social technologies into the different stages of the business process life cycle, in order to discover the hidden relationships among process artifacts. Social BPM takes advantage of social media tools, e.g., enriching business processes life cycle with tagging information [307], to improve communication. A novel research direction in advanced analysis techniques could be using natural language processing (NLP[7]) and coreference resolution [62] (CR) techniques to analyze the process-related documents and to discover more insight from hidden information in the text documents.

5.3 Business Data Analytics and Process Spaces

Existing business process management tools enable monitoring and analysis of *operational* business processes, i.e., the ones that are explicitly defined and the process is managed by a process-aware system, such as a workflow management system (WfMS) [7, 179, 289, 450]. However, in reality, only a fraction of process executions is supported by a WfMS, and business process is implemented across several heterogeneous IT systems. Gartner identifies process analysis and monitoring in such environments as a major challenge that plays a vital role in the survival and competitiveness of process management systems vendors in the near future [174].

As a motivating example, consider the processing of purchase orders in a typical enterprise depicted in Fig. 5.1: the order is first received by the company through a B2B hub, which can be either an in-house-developed Web service, an EDI receptor, or an e-commerce application such as Ariba, which begins with logging the related events and order verification. Once verified, the order is routed to the purchase order management system that initiates the approval process. Once approved, the processing of the order may require interaction with the workflow system, for procurement, but is also inevitably characterized by email and document exchanges among people as part of processing. At the same time, a notification may be sent to the invoice and payment systems and finally the shipping system to arrange for the shipping. During this process, documents (e.g., the purchase order and approval documents) may be stored in a document management system (e.g., a Microsoft SharePoint server) to facilitate collaborative editing and viewing of documents.

The above description shows that process execution information ends up being composed of a sea of apparently uncorrelated information items scattered across various systems in the enterprise. In fact, the situations where data on the entire process is conveniently located into one system and format do not happen in reality. Therefore, we need to analyze process data and understand the origin of data (i.e., *Provenance* [112]) in the enterprise so that they can be seen "as if" they were captured by a single business process management system. This implies to define

---

[7]http://nlp.stanford.edu/

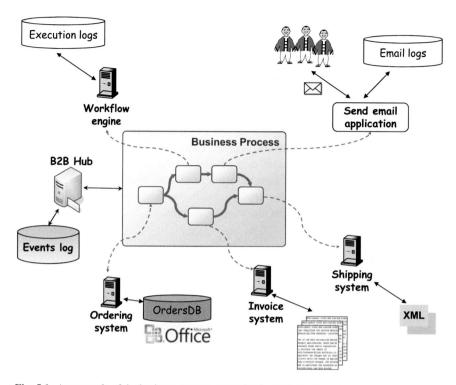

Fig. 5.1 An example of the business process execution in modern enterprises

a holistic view of the process executions over various information systems and services (i.e., process space) and to enable interpretation of the information in the enterprise in the context of process mining.

5.3.1 Process Space

A process space is essentially composed of (a) the definition of criteria or mechanisms for deciding which information items in the enterprise are *correlated*, i.e., belonging to the same execution (instance) of a process, (b) a way of *mapping* information items to process progression events (start and completion of process tasks), and (c) a *process model* of the processes in the enterprise. In a process space, different process models, mappings, and correlations can be defined over the same set of information items as different analysts may be interested in different views over such events (called as *process views*). For example, the shipments of a set of goods may be related from the view of the warehouse manager, but if the goods are the results of different orders, they are unrelated from the view of the sales manager. As an example scenario, consider ACME enterprise, depicted in Fig. 5.1, which supplies semiconductors to a variety of manufacturers in markets.

Example 1 (Search and Query Process Spaces) A business manager, Amy, receives a complaint email about a purchase order number 325 from a manufacturer. To properly respond to the complaint, she wants to find all the information on the process execution from the order request to the order completion, including product order details, people involved in the order approval process, emails and documents exchanged among them, payment and billing information, shipping information, etc. For instance, these include email ABC, shipping document PO325.xml, and invoice xyz.pdf. Currently, Amy has to *individually* search each data sources of the enterprise and look for information of the order number 325. For instance, she first needs to search the order DB, and get the invoice number from there. Then she has to search the invoice system to find related information and also payments and shipping subsequently going through emails and documents stored in various systems. This is a very daunting and time-consuming task.

A dataspace management system [168], which aims at making it possible to access data from multiple and heterogeneous data sources, does not provide the ability to interpret the data in the context of process executions. Other available techniques and technologies, e.g., enterprise information integration and enterprise search tools [205], also do not provide ways to find information regarding process executions in the enterprise. What is needed here is a kind of a process space management system (PSMS) that can offer the functionalities required in this scenario including: (a) allowing to browse all the information about process executions across data sources and to identify relationships among information in terms of process execution and (b) enabling to index a large variety of information including business documents (e.g., as Microsoft Word, emails, tuples in DB and XML documents) along with process execution context (e.g., events or process instances) in order to support efficient search and query of process executions.

Example 2 (Monitoring and Analyzing Process Spaces) An IT manager, Bob, is interested in monitoring and analyzing process executions to prevent similar complaints from happening in the future. He may ask which purchase order cannot be completed at the scheduled date, where the bottlenecks in the process are, and who has been working on the order. In addition, he wants critical situations to be identified and alerted before such situations arise, e.g., when there is a high probability that the shipping phase is not going to finish on time [102]. With current technologies such as WfMS and BPI techniques [196], it is difficult to monitor and analyze process executions supported by one or more systems (Web sites, databases, document management systems, ERP systems, message brokers, workflow systems, etc.), which not all of them are process aware.

Current process monitoring and analysis approaches (e.g., [51, 53]) also require a process model of the process to be analyzed. Hence, this limits the analysis to workflow log meaning that we can only monitor and analyze a fraction of processes in an enterprise. An important enabler of process analysis across such systems is the ability to correlate events and documents and discover process views for various users including Bob. What is missing is such a system to enable performing analysis tasks on most or all sorts of processes not only those supported by the workflow systems. For example, the approach proposed in [195, 196] considered the existence of a process model and focused on mining process execution data using decision trees to find interesting correlations between process data and behaviors (e.g., delays, outcomes, and more generally, behavior patterns defined by analysts).

5.3.2   *Logical Components of Process Spaces*

Business processes in an enterprise are implemented using various (heterogeneous) information systems and services. A *process space* is the set of data sources containing information related to the execution of processes in the enterprise over which we superimpose a business process metaphor. Some examples of process spaces are purchase order process space, insurance claim process space, mortgage application process space, and auction process space. The data sources in the process space can be categorized into two types:

- **Event data sources** refer to data sources that store metadata about the *events* related to the execution of business process and exchange of business documents and messages between information systems and services in the enterprise. The metadata include information such as time stamp and sender and receiver of documents or messages. Events may be recorded by various logging systems, e.g., workflow logs, Web service logs, Web logs, and email logs (see Fig. 5.1). Examples of such systems are included in SAP NetWeaver, IBM WebSphere, and HP SOA Manager. The level of details of information recorded in logs varies with logging systems.
- **Business data sources** refer to data sources that contain the business documents and messages that are exchanged or task execution data that are produced during business process executions. This data may be stored as files of various formats in file systems and repositories, as tuples in the database, or as text in Email systems (see Fig. 5.1 for some examples).

These two types of data sources are maintained separately or in some cases using the same systems. For instance, B2B hub may record metadata about the exchange of documents, but not the actual documents that are exchanged, and the actual documents may be stored in a document management system. In other cases, the event related to the exchange of an XML message between Web services along with the message may be stored in the same service log.

The superimposition of a process metaphor over the data sources requires:

(a) If the event data sources and business data sources are separate, identification of the *correspondence* business documents and event data. This is because neither of this information may be sufficient by itself to allow interpretation of data in terms of process execution. An *information item* refers to an event and its correspondent business data.

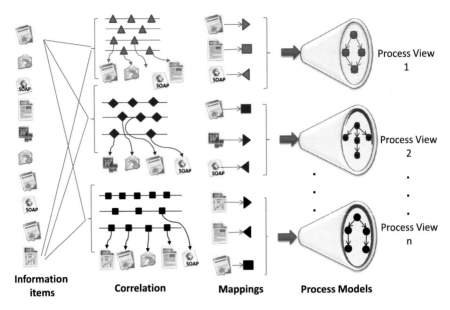

Fig. 5.2 Process views defined over information items characterize a process space

(b) *Mappings* of information items to the progression of process tasks (start, execution, and completion of process tasks). A *process item* refers to an information item mapped to a process task.
(c) Definition of criteria or mechanisms to *correlate* process items into process instances. A process instance represents the tasks performed during a process execution to achieve a business goal.
(d) The *model* of the underlying process followed by the process instances. This can be used as a reference for asking queries.

Therefore, the logical components of a dataspace are data sources, correspondences, mappings, correlations, and process models. As mentioned in the introduction, different mappings, correlations, and process models can be defined on the same set of information items. This is not only because the underlying information may belong to different processes but also because different users (analysts) look at the same data from different perspectives. For instance, consider information related to purchase orders *PO1, PO2, PO3, PO4, PO5*, and *PO6*, which belong to six different process instances from the perspective of a purchase order manager. However, if *PO1* and *PO2* are shipped together and *PO3, PO4, PO5*, and *PO6* together, then they belong to only two different instances from the warehouse manager's perspective. A *process view* refers to a given way of mapping, correlation, and corresponding process model. Figure 5.2 shows process views defined in a process space starting from the information items. In other words, It is also possible to characterize a process space using process views defined over information items.

A process view may be nested. For instance, the process view of the purchase order management system is nested within that of the whole enterprise, considered as a subprocess. This allows to look at a process space at various levels of abstractions and granularities from the high level (enterprise level) to details of a process execution. This covers the needs of various users in the enterprise.

5.3.3   Process Space Management System

PSMS [344] is proposed to enable interpretation of information in the process space. A PSMS offers the following "typical" categories of functionalities: *process space definition/discovery*, *process space analysis*, and *end-user tools for process space exploration and visualization*. Figure 5.3 shows an example of an architecture for a PSMS organized in three layers on top of data resources in a process space.

Process Space Definition/Discovery The main step toward development of a PSMS is to define the process space that is to identify the logical components of the process space. This can be done both by human users to manually define or automatically discovered from the data sources in the process space. Therefore, the PSMS should allow users to manually define, update, and maintain information

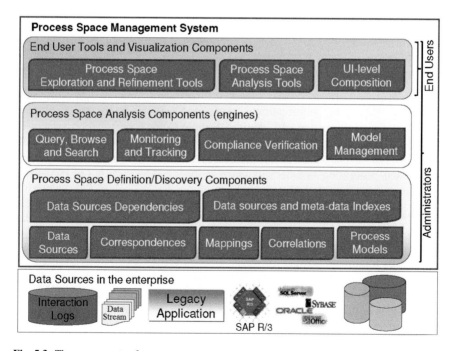

Fig. 5.3 The components of a process space management system

about these components. It also should provide automated support for discovery of the components. In particular, it must support incremental discovery and evolution of the process space components as more information become available in the enterprise.

In order to facilitate this task, the system should maintain a catalog of the metadata about each component of the process space. For instance, for a data source this information includes its name, location, owner, and the names of processes that their information can be found in the data source. Similarly, the catalog may contain information about the correspondences including evidences that the events and business data are related. Other information in the catalog includes a description of criteria used for correlation of process items and also mappings. In addition, the method on how that information is created, its creator, and the degree of confidence in the information, if they are discovered automatically, have to be stored.

In addition to the logical components of the process space, there are two other first-class objects in the process space that are used in many other higher-layer components, which are described in the following:

Indexes Indexes can also be considered as first-class objects in the process space. Indexes may be created on all process-related information on data sources and more importantly on the stored metadata about the process space logical components. The reason is that indexes support performing efficient searching, querying, and analysis of the process space. Indexes can be seen as summary information about the process space that can be used to answer many queries without the need to (re)compute them from the process space. This component should be adaptive, i.e., allow to update the indexes as the data sources become (un)available. The indexing mechanism may allow accessing the same process item via multiple references to it, e.g., by process instance, by process name, by the method of correlation, etc.

Dependencies Another first-class object in the process space is the relationship (dependencies) among information systems and services in terms of process execution. This information can identify the role of each information system in the process and also on systems that it depends upon to function. This information also allows to have a profile of information systems to know, e.g., which systems cooperate in processing an order and which services can be impacted by the failure of a service or system.

Process Space Analysis Components As depicted in Fig. 5.3, a PSMS offers several interrelated components for analyzing, querying, and monitoring of process executions in process spaces, some of which are generations of components provided by traditional WfMSs. However, unlike a WfMS which owns and controls all the process-related data, a PSMS allows the information to be independently managed by various information systems, and it provides a new set of services over the aggregate of these systems. In particular, the following analysis tasks are desired to be supported: (a) browse, query, and search engine, (b) monitoring and tracking engine, (c) compliance verification engine, (d) compliance verification engine, (e) model management engine, and (f) process space exploration and visualization tools.

Browse, Query, and Search Engine This component helps in browsing, querying, and searching the process space. The information in the catalog can be used to browse different information systems and repositories (e.g., logs and document management systems) that contain information on process executions and understand the correspondences, mappings, and correlations between them. Another functionality of this component is enabling querying of logs and documents. This includes queries that uses data sources from different systems (e.g., SOAP messages, logs, and emails related to the processing of the same order). An example of such queries is *select all orders approved by John where the time elapsed between order receipt and order shipment is more than 20 days.* Note that not all data sources may support the query of the same level of expressiveness, and also they may not support queries with process information. In this case, a PSMS may extend these data sources to make them process aware or to translate the queries into appropriate queries for such data sources.

Performing OLAP-style analysis on executions is also another type of queries. For instance, queries such as *how many people on average touch an order? what is the overhead and delays caused by human interaction vs machines? Which factors contribute to a process execution falling below the desired quality targets?*; In addition, the system should be able to search the content of documents and event logs for process-related information, e.g., based on keywords. Examples of such search queries include *find all purchase orders in which CPUs are requested.* As process space is potentially huge, it is important that the system can help users in formulating queries, e.g., by providing hints, identifying (non-)plausible queries, and proposing visualized approaches for query formulation.

The searching and querying functionality should allow posing queries at various levels of abstractions (e.g., at the enterprise level or at the process execution level). The search on the metadata about the process space, e.g., on the data sources, correspondences, mappings, correlations, and even process models, is another type of required querying and search.

Monitoring and Tracking Engine Monitoring of process execution and tracking the progress of a given instance is one of the important functionalities required by users in the enterprise. Examples of questions that could be answered by monitoring include *which route did a given order take? where did it get stuck? who has been working on it? where was the bottleneck? has it been shipped? and has it been paid?* To enable answering the above questions, the system enables accessing to up-to-date information about the process execution. The monitoring may also be performed at various levels of abstractions.

Compliance Verification Engine This component enables to verify compliance of process executions in the process space with policies and regulations. For instance, we may be interested to know if *privacy policies are respected by the way information are processed in the process space.*

Model Management Engine This component provides operators for analysis and management of process models at various levels of abstraction. Examples of operators include identifying *subsumption, replaceability, compatibility,* and *part-of* relationships between process models. These are helpful to understand if the model which followed in the process space complies to the one designed in the enterprise, and also the relationship between processes executed by different systems, and also if there is any overlap between them.

End-User Tools and Visualization Components   Providing tools and visualization techniques to explore the dataspace and perform analysis tasks are required for end users. In particular, the following supports are useful in PSMS:

Process Space Exploration and Visualization Tools   These include tools that provide visual assistance to process space administrators in discovery process space components and other first-class objects and also refining and maintenance of metadata about these components. This may be provided by integrating or extending existing successful visualization and user interface paradigm such as spreadsheets to perform light process-related analysis tasks for end users. Spreadsheets are widely used end-user tools due to their simplicity, flexibility, and relative richness of analysis. Many vendors have already integrated spreadsheets into workflow systems such as OracleBI [367] and SAP BI [412]. The intention here is to take this to a higher level by making spreadsheet-like environments available for business process analysis and management in process spaces for end users.

Process Analysis Tools   These components provide user interface for performing the analysis operations offered by the components in the lower level. In particular, the provided interfaces may offer graphical environments to perform BPI tasks such as process monitoring; performance evaluation and measurements, e.g., in terms of KPI measures; and process verification and offer model management tools to use the engines over the process space.

User Interface-Level Composition   Once users have known the resources available in their process space, in terms of process execution, they may be able to build new applications by reusing existing applications. End users would appreciate tools that allow them to perform composition at the user interface level. Hence, this component should allow users to compose the functionality of various information systems in the process space and build mash-ups, e.g., using visual components such as pipes and widgets.

Using the current technology to achieve the functionalities that should be offered by a PSMS requires extensive (process-specific) development effort. This may require performing ad hoc, tedious, and error-prone tasks as discussed briefly earlier. The first step in building a process space is to identify its logical components. Identification can be achieved through manual definition of these components in the system and also by discovering information about them through automated approaches. Querying, browsing, and searching process spaces require having appropriate process analysis languages. Using existing technologies, we would have to resort to SQL queries and to ad hoc programs to extract the information we want from the data. However, the abstraction level of the information we want to extract is different from that of the analysis language (SQL). What is needed here is a way to ask high-level "process queries" or to perform process browsing, possibly with the same level of simplicity that IT managers are used to when dealing with reporting on Excel or in Web search. Visual query languages for business processes have been proposed [51].

In traditional data integration approaches (e.g., for building data warehouses or federated databases), data cleaning (consisting of detecting and correcting errors and inconsistencies from data) is an important step to improve the quality of data. Data of poor quality might cause significant cost escalation and time delays on business processing that relies on the data, e.g., mailing cost increase (posting mails to wrong addresses) and delivery delays (due to wrong addresses). Like data warehousing or decision support systems, the problem of cleaning process data (including events) is also one of the major problems in the area of process spaces. Data quality in process spaces is affected by the following reasons: (a) as in traditional data integration, data entry errors, inconsistent conventions, and poor integrity constraints and (b) bugs and flaws in the implementation of logging systems or the business process, exceptions in the process execution, and abnormal termination of the process interactions. Such data imperfection could lead to incorrect results in analysis results, erroneous process space discovery, inaccurate data correlation, etc. Hence, it is required to develop a data cleaning approach that effectively identifies the characteristics of process execution data and resolves the differences or conflicts of the data.

A process space must cover a wide range of heterogeneous data sources and applications. One major issue to accessing such systems is the inherent uncertainty about the correctness of information and the availability of some desired services over all data sources due to issues in the integration of heterogeneous data sources such as the level of supported interoperation, the expressiveness of supported query languages, noise, and various performances (e.g., legacy systems) that may affect the response time of the system. So a PSMS may be able to provide its service with a variable level of guarantees on different data sources. Indeed, achieving the same level of guarantees that a WfMS offers may not be possible. The research challenge is then how to define realizable, practical, and meaningful levels of service guarantees and defining existing trade-offs and factors which influence the quality and performance measures. Another related research challenge is how to make a PSMS robust, i.e., tolerant to the inaccuracies in the data sources and to follow the "best-effort" model in returning results of analysis.

5.4 Process Mining

In order to analyze process execution data, querying execution logs of completed business processes (i.e., process mining [2, 6]) received continuous attention in research. The goal of process mining is to simplify process queries and to semiautomate the query formation in order to easily establish links between the actual processes, their data, and the process models. Process mining subsumes process analytics and enhances it with more comprehensive insights on the process execution: process mining techniques can be used to identify bottlenecks and critical points through *replaying* the execution traces used to discover a process model and *enrich* the discovered model with quantitative information.

In particular, process mining helps in discovering and improving real processes by extracting knowledge from event logs through using process modeling/analysis, machine learning, and data mining techniques. The main concern of these approaches is to reverse engineer the definitions of business process models from execution logs of information system components. Moreover, depending on how much details the log gives, they can provide statistics about many aspects of the business processes such as: the average duration of process instances or the average resource consumptions.

Recently, the IEEE Task Force on Process Mining released a manifesto describing guiding principles and challenges in process mining [11], where the goal is to increase the maturity of process mining as a new tool to improve the (re)design, control, and support of operational business processes. In particular, process mining challenges include [2, 5, 6, 11]:

- Mining hidden and duplicate tasks: one of the basic assumptions of process mining is that each event is registered in the log. Consequently, it is challenging to find information about tasks that are not recorded. Moreover, the presence of duplicate tasks is related to hidden tasks and refers to the situation that one can have a process model with two nodes referring to the same task.
- Loops: in a process it may be possible to execute the same task multiple times, i.e., this typically refers to a loop in the corresponding process model.
- Temporal properties: the temporal metadata (e.g., event time stamps) can be used for adding time information to the process model or to improve the quality of the discovered process model.
- Mining different perspectives: understanding process logs in terms of its scope and details is challenging especially as it is subjective, depending on the perspective of the process analyst.
- Dealing with noise and incompleteness: the log may contain noise (e.g., incorrectly logged information) and can be incomplete (e.g., the log does not contain sufficient information to derive the process).
- Gathering data from heterogeneous sources: in modern enterprises, information about process execution is scattered across several systems and data sources.
- Visualization techniques: it helps in presenting the results of process mining in a way that people actually gain insight in the process.
- Delta analysis: it is used to compare the two process models and explain the differences. It can be useful as process models can be descriptive or normative.

As illustrated in Fig. 5.4, three types of process mining are recognized:

(a) Discovery: this technique takes an event log and produces a model without using any a priori information. For example, the α-algorithm [8] takes an event log and produces a Petri net [352] model which explains the behavior recorded in the log.

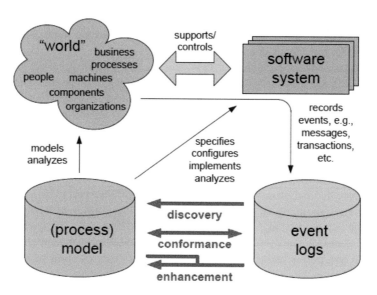

Fig. 5.4 Positioning of the three main types of process mining: (**a**) discovery, (**b**) conformance checking, and (**c**) enhancement [2]

(b) Conformance: in this technique an existing process model is compared with an event log of the same process. Conformance checking can be used to check if reality, as recorded in the log, conforms to the model and vice versa. For example, the conformance checking algorithm proposed in [402] can be used to quantify and diagnose deviations.

(c) Enhancement: this technique can be used to extend or improve an existing process model using information about the actual process recorded in some event log. In this context, two types of enhancement are recognized: repair, which can be used to modify the model to better reflect reality, and extension, which can be used to add a new perspective to the process model by cross correlating it with the log.

A recent book [2] and surveys [3, 5, 6] discuss the entire process mining spectrum from process discovery to operational support. Moreover, a growing number of software vendors added process mining functionality to their tools. For example, ProM [10] offers a wide range of tools related to process mining and process analysis. In particular, ProM is a workflow discovery prototype tool that offers some of above approaches. Agrawal et al. [21] proposed an approach to apply process mining in the context of WfMSs and to address the problem of model construction. Datta [127] proposed algorithms for the discovery of business process models. Also, similar approaches in the context of software engineering processes have been addressed in [122]. Herbst [224] presented a learning algorithm that is capable of inducing concurrent workflow models. The proposed approach focused on processes containing duplicate tasks and presented a specialization-based technique for discovering sequential model of process logs represented using HMM (hidden Markov model).

Conformance checking techniques [17, 351] are used to relate events in the log to activities in the model. Adriansyah et al. [17] presented techniques to measure the conformance of an event log for a given process model. The approach quantifies conformance and provides intuitive diagnostics and has been implemented in the ProM framework. Munoz-Gama et al. [351] presented an approach to enrich the process conformance analysis for the precision dimension. Some other examples of approaches focused on precision for: measuring the percentage of potential traces in the process model that are in the log [193], comparing two models and a log to see how much of the first model's behavior is covered by the second [323], comparing the behavioral similarity of two models without a log [148], and using minimal description length to evaluate the quality of the model [95].

Enhancement techniques heavily rely on the relationship between elements in the model and events in the log. These relationships may be used to: (a) replay the event log on the model, e.g., bottlenecks can be identified by replaying an event log on a process model while examining the time stamps [2], (b) modify the model to better reflect reality, and (c) add a new perspective to the process model by cross correlating it with the log. Subramanian et al. [429] proposed an approach for enhancing BPEL engines with facilities that permit satisfying self-healing requirements. Moreover, the concept of self-healing as a part of autonomic computing has been proposed in [258], where self-healing systems will automatically detect, diagnose, and repair localized problems resulting from failures. Diagnostic reasoning techniques and diagnosis-aware exception handlers for exception handling are proposed in [35]. Also, a framework for providing a proxy-based solution to BPEL, as an approach for dynamic adaptation of composite Web services, is presented in [164].

Process Mining vs. Data Mining Although process mining and data mining have lots of aspects in common, there are also fundamental differences in what they do and where they can be useful. The goal of data mining is to discover previously unknown interesting patterns in datasets. In this context, various methods at the intersection of database techniques such as spatial indices, artificial intelligence, machine learning, and statistics can be used. Data mining provides valuable insights through analysis of data and does not have concerns about the processes. This is where process mining provides the opportunity to get the same benefits of data mining, when working with processes and focusing on process improvements. In this context, unlike data mining, process mining focuses on the process perspective to find process relationships in the data. More specifically, process mining's perspective is not on patterns in the data but in the processes the data represents. Process mining can be seen as the "missing link" between data mining and traditional business process management. From the similarity point of view, both process mining and data mining use the mining techniques to analyze large volumes of data, e.g., process logs in process mining and electronic health records (EHR) in data mining. Moreover, both techniques produce information that can be helpful for making business decisions.

5.5 Analyzing Crosscutting Aspects in Processes' Data

Modern business processes have flexible underlying process definition where the control flow between activities cannot be modeled in advance but simply occurs during runtime [149]. The semi-structured nature of such process's data requires analyzing process-related entities, such as people and artifacts, and also the relationships among them. In many cases, however, process artifacts evolve over time, as they pass through the business's operations. Consequently, identifying the interactions among people and artifacts over time becomes challenging and requires analyzing the crosscutting aspects [156] of process artifacts. In particular, process artifacts, like code, have crosscutting aspects such as versioning (what are the various versions of an artifact, during its life cycle, and how they are related) and provenance [112] (what manipulations were performed on the artifact to get it to this point).

The specific notion of business artifact was first introduced in [358] and was further studied, from both practical and theoretical perspectives [80, 180, 231]. However, in a dynamic world, as business artifacts change over time, it is important to be able to get an artifact (and its provenance) at a certain point in time. It is challenging as annotations assigned to an artifact (or its versions) today may no longer be relevant to the future representation of that artifact: artifacts are very likely to have different states over time and the temporal annotations may or may not apply to these evolving states. Consequently, analyzing evolving aspects of artifacts (i.e., versioning and provenance) over time is important and will expose many hidden information among entities in process data. This information can be used to detect the actual processing behavior and, therefore, to improve the ad hoc processes.

As an example, knowledge-intensive processes, e.g., those in domains such as healthcare and governance, involve human judgments in the selection of activities that are performed. Activities of knowledge workers in knowledge-intensive processes involve directly working on and manipulating artifacts to the extent that these activities can be considered as artifact-centric activities. Such processes almost always involve the collection and presentation of a diverse set of artifacts, where artifacts are developed and changed gradually over a long period of time. Case management [433], also known as case handling, is a common approach to support knowledge-intensive processes. In order to represent crosscutting aspects in processes, there is a need to collect metadata about entities (e.g., artifacts, activities on top of artifacts, and related actors) and relationships among them from various systems/departments over time, where there is no central system to capture such activities at different systems/departments.

Many approaches [81, 118, 180, 231] used business artifacts that combine data and process in a holistic manner and as the basic building block. Some of these works [180, 231] used a variant of finite-state machines to specify life cycles. A new line of works, such as [60, 150], considers an artifact-centric activity model for business processes. These models support timed queries and enable weaving crosscutting aspects, e.g., versioning and provenance, around business artifacts to imbue the artifacts with additional semantics that must be observed in constraint and querying ad hoc processes.

5.6 Provenance and Evolution of Business Artifacts

Provenance refers to the documented history of an object (e.g., documents, data, and resources) or the documentation of processes in an object's life cycle, which tracks the steps by which the object evolved and was derived [112]. This documentation (often represented as graphs) should include all the information necessary to reproduce a certain piece of data or the process that led to that data [342]. The ability to analyze provenance data is important as it offers the means to verify business data products, to infer their quality, and to decide whether they can be trusted [341]. In a dynamic world, as data changes, it is important to be able to get a piece of data, as it was, and its provenance data, at a certain point in time. Under this perspective, the provenance queries may provide different results for queries looking at different points in time. In this context, enabling time-aware querying of provenance information is challenging and requires explicitly representing the time information in the provenance graphs and also providing abstractions and efficient mechanisms for time-aware querying of provenance graphs over an ever-increasing volume of data.

Prior work on modeling and representing provenance metadata [112, 170, 421] (e.g., lineage, where-provenance, why-provenance, dependency-provenance, how-provenance, and provenance-trace models) models provenance as a directed acyclic graph, where the focus is on modeling the process that led to a piece of data. They present vocabularies to model process activities and their causal dependency, i.e., the relationship between an activity (the cause) and a second activity (the effect) where the second activity is understood as a consequence of the first. In a dynamic world, data changes, so the graphs representing data provenance evolve over time. It is important to be able to reproduce a piece of data or the process that led to that data for a specific point in time. This requires modeling *time* as a first-class citizen in the provenance models. Times, intervals, and versioning can be very important in understanding provenance graphs as the structure of such graphs evolves over time. Today's approaches in modeling provenance, e.g., in OPM, treat time as a second-class citizen. Considering time as a first-class citizen will enable retrieving multiple snapshots of entities (versions) in the past which can help in capturing the provenance for each version of an entity independently. Moreover, it can help in understanding the role of each entity in the temporal context of the entire system.

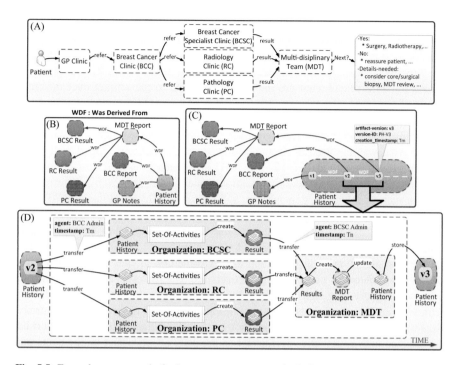

Fig. 5.5 Example case scenario for breast cancer treatment including a case instance (**a**); parent artifacts, i.e., ancestors, for a patient history document (**b**) and its versions (**c**); and a set of activities which shows how version v_2 of a patient history document develops and changes gradually over time and evolves into version v_3 (**d**)

To understand the provenance, consider a scenario based on breast cancer treatment cases in Velindre hospital [433]. Figure 5.5a represents a case instance, in this scenario, where a general practitioner (GP), suspecting a patient has cancer, updates patient history and refers the patient to a breast cancer clinic (BCC). BCC checks the patient's history and requests assessments such as an examination, imaging, fine-needle aspiration, and core biopsy. Therefore, the BCC administrator refers the patient to the breast cancer specialist clinic (BCSC), radiology clinic (RC), and pathology clinic (PC), where these departments apply medical examinations and send the results to the multidisciplinary team (MDT). The results are gathered by the MDT coordinator and discussed at the MDT meeting involving a surgeon oncologist, radiologist, pathologist, clinical and medical oncologist, and nurse.

Analyzing the results and the patient history, MDT will decide for the next steps, e.g., in case of positive findings, nonsurgical (radiotherapy, chemotherapy, endocrine therapy, biological therapy, or bisphosphonates) and/or surgical options will be considered. During the interaction among different systems, organizations, and care team professionals, a set of artifacts will be generated. Figure 5.5b represents parent artifacts, i.e., ancestors, for a patient history document, and Fig. 5.5c represents parent artifacts for its versions. Figure 5.5d represents a set of activities which shows how version v_2 of a patient history document develops and changes gradually over time and evolves into version v_3.

Considering this scenario, modeling and analyzing provenance data will enable the process analyst to answer the following questions: who was involved in generating an artifact? what are the changes applied to the artifact over different points of time and who was involved in these processes? how one version of the artifact evolved from another version? what was the used artifact and the purpose to obtain result? what was the used artifact and the collected data used to obtain result? what was the used artifacts and the security processes applied to it?, etc. Several provenance models [112, 170, 421] have been presented in a number of domains (e.g., databases, scientific workflows, and the semantic Web), motivated by notions such as influence, dependence, and causality. Why-provenance [170, 421] models the influences that a source data had on the existence of the data. Where-provenance [112] focuses on the dependency to the location(s) in the source data from which the data was extracted. How-provenance [112, 170] represents the causality of an action or series of actions performed on or caused by source data. Discovering historical paths through provenance graphs forms the basis of many provenance query languages' queries [226, 253]. Temporal databases [338] enable retrieving multiple snapshots (versions) of data artifacts at different points in time. However, a temporal database does not capture important information for data provenance such as activities performed on the data, agents acting on the data, and the relationships that the different versions of artifacts have for each other in various points in time. Approaches for modeling and querying graphs (e.g., [33, 56, 189]) can be used for querying provenance data.

Besides provenance and versioning, other aspects of business artifacts such as security (who has access to the artifact over time), privacy (what actions were performed to protect or release artifact information over time), and trust (the credibility of users and their posted and shared content in a particular domain) need to be analyzed. Analyzing these aspects will expose many hidden interactions among entities in process graphs. For example, in current outsourcing practices, clients usually focus primarily on business objectives, and security is negotiated only for communication links. In such scenarios, strong protection of a communication link is of little value if data can be easily stolen or corrupted while on a supplier's server. For example, analyzing such manipulated data (e.g., stolen or unauthorized accessed data) may lead to unreliable decisions.

Chapter 6
Tools, Use Cases, and Discussions

The continuous demand for the business process improvement and excellence has prompted the need for business process analysis in the enterprise. Recently, the business world has begun to become increasingly dynamic as various technologies such as the Internet and email have made dynamic processes more prevalent. Following this, the problem of understanding business process execution has become a priority in rapidly changing and knowledge-intensive organizations. In particular, analyzing business process execution is a crucial requirement for many end users in order to monitor, analyze, understand, and improve their business. This chapter provides an overview of open-source and commercial software for process analytics. The software for process analytics can be applied to the rich source of events, which document the execution of processes and activities within BPM systems, in order to support decision-making in organizations. We provide a summary and comparison of existing open-source and commercial software for process analytics, including real-world use case scenarios, followed by a discussion and future directions on some of the emerging and hot trends in the business process management area such as process spaces, big data for processes, crowdsourcing, social BPM, and process management on the cloud. We briefly describe the core essence of these directions and discuss how they can facilitate the analysis of business processes.

6.1 Observations

Understanding, analyzing, and ultimately improving business processes are the goals of enterprises today. Most related work in the area of analyzing business process execution assumes well-defined processes; however, the business world is getting increasingly dynamic, and there are cases where the process execution path can change in a dynamic and ad hoc manner. The current state of the art in querying business processes does not provide sufficient techniques for the analysis

© Springer International Publishing Switzerland 2016
S.-M.-R. Beheshti et al., *Process Analytics*, DOI 10.1007/978-3-319-25037-3_6

of process data. For example, some of the basic assumptions of existing BP querying techniques are that each event should be registered in the log, the BP models should be predefined and available, and the execution traces should comply with the defined process models.

In particular, the understanding of business processes and analyzing BP execution data are difficult due to the lack of documentation and especially as the process scope and how process events across these systems are correlated into process instances are subjective, depending on the perspective of the process analyst. Consequently, there is a need for querying approaches that enables analysts to analyze the process events from their perspectives, for the specific goal that they have in mind, and in an explorative manner.

Moreover, most objects and data in the integrated process logs are interconnected, forming complex, heterogeneous but often semi-structured networks and can be modeled using graphs. Understanding modern business processes entails identifying the relationships among entities in process graphs. Viewing process logs as a network and process graphs and studying systematically the methods for mining such networks, of events, actors, and process artifacts, are a promising frontier in database and data mining research: process mining provides an important bridge between data mining and business process modeling and analysis [11].

There are many studies on the analysis of graphs, such as network measures [463], statistical behavior study [355], modeling of trend and dynamic and temporal evolution of networks [228, 268], clustering [387], ranking [430], and similarity search. All these approaches can be leveraged for mining and analyzing process graphs. Moreover, for effective discovery of ad hoc process knowledge, it is important to enhance process data by various data mining methods, i.e., to help data cleaning/integration, trustworthiness analysis, role discovery, and ontology discovery, which in turn help improving business processes.

To address these challenges, a set of works [346, 403] focused on the correlation discovery between events in process logs, i.e., event correlation is the process of finding relationships between events that belong to the same process execution instance. In particular, the problem of event correlation can be seen as related to that of discovering functional dependency [55, 299, 380] in databases. These works are complementary to process mining techniques as they enable grouping events in the log into process instances that are then input to process mining algorithms.

Some other related works focused on converting process execution data into knowledge to support the decision-making process [38, 265]. They presented a family of methods and tools for developing new insights and understanding of business performance based on collection, organization, analysis, interpretation, and presentation of ad hoc process data. While existing analytics solutions, e.g., OLAP techniques and tools, do a great job in collecting data and providing answers on known questions, key business insights remain hidden in the interactions among objects and data. Existing approaches [111, 162, 214, 248, 262, 388], in online analytical processing (OLAP) on graphs, took the first step by supporting multidimensional and multilevel queries on graphs; however, much work needs to be done to make OLAP heterogeneous networks a reality [215].

Another related work [56, 57, 60] focused on providing a framework, simple abstractions, and a language for the explorative querying and understanding of process graphs from various user perspectives. The framework caters for life cycle activities important for a wide range of processes, from unstructured to structured, including understanding, analyzing, correlating, querying, and exploring process execution data in an interactive manner. Moreover, the framework provides techniques for applying existing mining algorithms and analytics to process data.

6.2 Open-Source and Commercial Tools for Process Analytics

Applications of process analytics extend to nearly all managerial functions in organizations. For example, considering financial services, applying analytics on customer dossiers and financial reports can specify the performance of the company over periods of time. Such operations require the use of various open-source and commercial software for process analytics. These software can be applied to the rich source of events that document the execution of processes and activities within BPM systems, in order to support decision-making in organizations. In this context, various existing tools focus on the behavior of completed processes, evaluate currently running process instances, or focus on predicting the behavior of process instances in the future. This section provides a summary and comparison of existing open-source and commercial software for process analytics.

Existing software for process analytics focus on two main perspectives: compliance and performance. Commercial process analytics tools such as IBM,[1] Oracle,[2] SAP,[3] and ARIS[4] software leverage a combination of process management/integration and business rule management software, to support decision-making in organizations. These tools provide: (a) frameworks for data modeling and metadata management, to support processes with data analytics; (b) support for human collaboration to improve process efficiency and quality; (c) support for monitoring processes, cases, tasks, and people to support real-time analytics; (d) support for cloud integration and API management, to empower analysts quickly and securely connect with the enterprise data while protecting against threats; and (e) analytics for workflows, tasks, processes, cases, actors, and artifacts.

For example, Oracle business process management suite[5] provides workflow/task analytics such as productivity indicators and task inflow/outflow with time series

---

[1] http://www-03.ibm.com/software/products/en/category/BPM-SOFTWARE

[2] www.oracle.com/us/technologies/bpm/suite/overview/index.html

[3] www.sap.com/pc/tech/business-process-management/software/overview.html

[4] www.softwareag.com/corporate/products/aris_alfabet/default.asp

[5] www.oracle.com/us/technologies/bpm/suite/overview/index.html

Table 6.1 Comparison of business process modeling tools

Name	Creator	Features
IBM Process Designer[a]	IBM	Modeler, simulation, execution
HP Process Automation[b]	HP	Automation, content management
ARIS Express[c]	Software AG	Modeler; supports for BPMN2[n] and event-driven process chain (EPC) process modeling and management
jBPMN[d]	NetBeans	Modeler
iServer[e]	Orbus Software	Modeler
Pega BPM[f]	Pegasystems	Modeler, simulation, execution
Enterprise Architect[g]	Sparx Systems	Modeler; supports for BPMN, BPEL, UML, and SysML modeling
SYDLE SEED[h]	SYDLE Systems	Modeler, simulation, execution
TIBCO ActiveMatrix[i]	TIBCO Inc.	Modeler, simulation, execution
Activiti Modeler[j]	Alfresco	Modeler, simulation, execution
AuraPortal[k]	AuraPortal	Modeler, execution, simulation
Bonita BPM[l]	Bonitasoft	Modeler, execution, simulation
IBM Rational System[m]	IBM	Modeler and querying support; RDF data model and SPARQL query

[a]http://www.ibm.com/support/docview.wss?uid=swg27023009
[b]http://www8.hp.com/us/en/software-solutions/business-process-automation/
[c]http://www.ariscommunity.com/aris-express
[d]http://plugins.netbeans.org/plugin/50735/jbpmn
[e]http://www.orbussoftware.com/products/iserver/
[f]http://www.pega.com/bpm
[g]http://www.sparxsystems.com/platforms/business_process_modeling.html
[h]http://www.sydle.com/bpms
[i]http://www.tibco.com/
[j]http://www.activiti.org/
[k]http://www.auraportal.com/
[l]http://www.bonitasoft.com/
[m]http://www-03.ibm.com/software/products/en/ratisystarch
[n]List of Active BPMN 2.0 Engines: https://en.wikipedia.org/wiki/List_of_BPMN_2.0_engines

to help with completed work analysis. It also provides process analytics, such as performance and workload metrics to detect bottleneck in processes, case milestone-based analysis, stakeholder-based analysis, activity-based analysis, case-time-based analysis, and artifact evolution analysis to provide process performance monitoring through either BPM Workspace dashboards or Oracle BAM. Table 6.1 represents a comparison of business process modeling tools.

6.2.1 BPM in the Cloud

Besides the need to extend decision support in process data analysis scenarios, the other challenge is the need for scalable analysis techniques. Similar to scalable data processing platforms [483], such analysis and querying methods should offer automatic parallelization and distribution of large-scale computations, combined with techniques that achieve high performance on large clusters, e.g., cloud-based infrastructure, and be designed to meet the challenges of process data representation that should capture the relationships among data (mainly, represented as graphs). In particular, there is a need for new scalable and process-aware methods for querying, exploration, and analysis of process data in the enterprise because: (a) process data analysis methods should be capable of processing and querying a large amount of data effectively and efficiently and therefore have to be able to scale well with the infrastructure's scale and (b) the querying methods need to enable users to express their data analysis and querying needs using process-aware abstractions rather than other lower-level abstractions.

Recently some vendors provide BPM in the cloud solutions to offer visibility and management of business processes, low start-up costs, and fast return on investment. These solutions can drive new growth opportunities, increase profit margins for the private sector, and achieve more efficient and effective missions for organizations. In particular, business process management tools in the cloud enable strategic process improvement, reduced technology cost, and better alignment of IT with business goals. Table 6.2 represents a comparison of BPM in the cloud tools.

Open-Source Software for Analyzing Big Process Data Large companies such as Facebook, Yahoo, Twitter, and LinkedIn benefit and contribute working on opensource projects. Big process data infrastructure can benefit from these open-source software such as: (a) Apache Hadoop [470], software for data-intensive distributed applications, based on the MapReduce programming model and a distributed file system (HDFS); (b) Apache Hadoop-related projects[6], Apache Pig, Apache Hive, Apache HBase, Apache ZooKeeper, Apache Cassandra, Cascading, Scribe, and many others; (c) Apache S4 [354], platform for processing continuous data streams; (d) Apache Spark, cluster computing framework with in-memory primitives that provide performance up to 100 times faster (compared to MapReduce) for certain applications; and (e) Storm[7], software for streaming data-intensive distributed applications.

Open-Source Software for Mining Big Process Data The big data problem can be seen as a massive number of small data islands from personal, shared, and business data. Mining this data is of high interest as leading companies such as Google, Apple, Facebook, Yahoo, and Twitter are starting to look carefully to this data to find useful patterns to improve user experience. In big data mining, there are

---

[6]http://www.apache.org/

[7]http://storm-project.net

Table 6.2 Comparison of BPM in the Cloud tools

Name	Creator	License	Features
IBM Blueworks[a]	IBM	Commercial	IBM BPM on Cloud is a BPM cloud service that includes tooling and environments to design, test, and run process applications, along with capabilities for monitoring and optimizing work that is run within the platform
Appian[b]	Appian	Commercial	The Appian BPM suite delivered in the cloud is a secure, scalable, and reliable way to deploy BPM solutions as a service in the cloud. Appian in the cloud has the same functionality as traditional on-premise BPM software deployments
Bizagi[c]	Bizagi	Commercial	Bizagi can be used to automate processes and has made available a set of executable process templates including Six Sigma process management, loans, insurance, transactional process, etc. It provides a set of widgets to add functionality to processes
Signavio[d]	Signavio	Commercial	Signavio Process Editor is a Web-based business process modeling tool that enables the creation of process diagrams using the business process model and notation, and it is available as software as a service (SaaS) and for on-premise installations
ProcessMate[e]	ProcessMate	Commercial	ProcessMate is a cloud software that helps keep track of processes; manages related documents, related data, and communication; and provides on-time notifications that help minimize delays
OpenText[f]	OpenText	Commercial	The OpenText BPM cloud suite offers a complete middleware platform as a service for BPM, integration, application development and enterprise mobility in the cloud
BonitaCloud[g]	Bonitasoft	Open source	Bonita BPM is an open-source business process management and workflow cloud-based suite. It allows the user to graphically modify business processes following the BPMN standard
jBPM[h]	Red Hat	Open source	BPM is a flexible business process management (BPM) suite. It makes the bridge between business analysts and developers. The core of jBPM is a lightweight, extensible workflow engine that allows analysts to execute business processes

(continued)

Table 6.2 (continued)

Name	Creator	License	Features
ProcessMaker[i]	ProcessMaker	Open source	ProcessMaker is a Web-based cost-effective open-source business process management (BPM) and workflow software application. It provides a drag-and-drop interface which makes it easy for business analysts to model approval-based workflows
Activiti[j]	Activiti Co.	Open source	Activiti is a cloud-based business process management (BPM) platform. It is open source and distributed under the Apache License. Activiti runs in any Java application, on a server, on a cluster, or in the cloud
Intalio[k]	Intalio Inc.	Open source	Intalio is a cloud-based business process management (BPM) platform. It provides a set of runtime components that support human workflow within a service-oriented architecture (SOA)
Camunda[l]	Camunda	Open source	Camunda is an open-source platform for workflow and business process automation. It executes BPMN 2.0, is very lightweight, and scales like hell. Camunda can be added to Java application as a library or run it as a container service in Tomcat, JBoss, GlassFish, IBM WAS, or Oracle WLS

[a]https://www.blueworkslive.com/
[b]http://www.appian.com/bpm-software/cloud-bpm/
[c]http://www.bizagi.com/
[d]http://www.signavio.com/
[e]http://processmate.net/
[f]http://www.opentext.com
[g]http://www.bonitasoft.com/
[h]http://www.jbpm.org/
[i]http://www.processmaker.com/
[j]http://activiti.org/
[k]http://www.intalio.com/
[l]http://camunda.org/

many open-source initiatives that can be reused in mining process-related data. The most popular are the following:

(a) Apache Mahout[8]: scalable machine learning and data mining open-source software based mainly on Hadoop

(b) R [389]: open-source programming language and software environment designed for statistical computing and visualization

---

[8]http://hadoop.apache.org

(c) MOA [82]: stream data mining open-source software to perform data mining in real time
(d) Vowpal Wabbit [280]: an open-source project started at Yahoo Research and continued at Microsoft Research to design a fast, scalable, useful learning algorithm
(e) Pegasus [250]: a big graph mining system built on top of MapReduce
(f) GraphLab [301]: a high-level graph-parallel system built without using MapReduce, and DREAM [212], a distributed graph engine with adaptive query planner

In the context of big data, there are some challenges in the crossover between process mining and data mining techniques. For example, process data is scattered across many resources and requires distributed mining; however, many data mining techniques are not trivial to paralyze, and consequently, there is a need for new approaches to support distributed process data mining with practical and theoretical analysis. More importantly, business data may be evolving over time [58], and it is important that the process-related data mining techniques be able to understand the changes in data over time and provide analytics for different periods of time.

6.2.2   Business Process Analytics: Practical Use Case Scenario

The following scenario describes a real-world scenario in a large service company, and let us call it Acme. Acme is offering services to large and small healthcare customers. They are working with a large healthcare provider, and let us refer to it as HealthAlliance, which is managing a large number of healthcare facilities. Each facility has its own operational systems, which collect large amounts of data related to the healthcare provision services. The systems capture data from the moment the patients check in, doctor visits, their labs, routing among different specialists, and in some cases referral to the hospitals, their hospital stay, and posthospital care. The HealthAlliance group has grown over several years through multiple mergers and acquisitions. Each new healthcare group comes with its own systems and processes. The integration with every new acquired provider has not been easy. It is during the last two acquisitions that they approached Acme to help them look at their healthcare process and integration process. They had the following problems and questions when approached Acme:

1. For their already integrated healthcare system, they have an understanding of how their healthcare process should work at different stages of the patient care life cycle. The question is whether the current process execution is compliant to the assumed process model potentially through looking at the data.
2. What are the bottlenecks in the current patient care process, and how can it be improved?
3. How to facilitate and simplify integration with newly acquired healthcare units in terms of systems and processes?

In the following, we review how the introduced suite of available process modeling, querying, and analytics helps addressing the above problems for HealthAlliance. As the starting point, the business process analyst team of Acme worked with the HealthAlliance healthcare process owner team to prepare an updated business process model using existing business process modeling tools, in particular IBM Blueworks (Table 6.2), as the team was regularly using it in their other projects. Also, as a cloud-based tool, the team could use it to collaboratively update the model. In order to understand how the current process works, the team relied on deploying a proprietary graph-based process discovery tool (equivalent to ProM [10]) for each of subsystems. Although not used in the context of this particular project, the team could use the process model matching techniques and tools to compare the designed process model and discovered process model based on the process mining tools.

One of the newly acquired healthcare units had a number of patient subsystems that were not connected to each other, though the information of patients that have visited those departments was present in those subsystems. In order to find out how these systems can be implemented, the correlation discovery method introduced in [346] was used to find rules that could be used to correlate process events related to patient records in different stages of the care process. In this process, for integrating the patient data record, schema matching techniques can be used (Sect. 3.2 in Chap. 3) to simplify the process of attribute mapping between different systems. Though, correlation discovery method would function without such a step, as well.

The other important question was how to find the bottlenecks in the current processes. Given there was no business process monitoring systems in place, the team could build a big data platform (based on the large process knowledge graph methods introduced in Sect. 5.2 in Chap. 5) to process all patient-related events in the systems for the HealthAlliance healthcare systems to build a knowledge graph that captures the relationships among different entities (healthcare team including doctors, nurses, accounting, etc.) and events from different referrals, visit schedules, visit report writings, lab test results, etc., and also be annotated with event information including the timing of the event. This graph then can be queried using annotated business process execution data using methods introduced in Chap. 4, such as BP-SPARQL [54, 56, 61], in order to answer specific questions on what are the tasks/events that create the most bottleneck in the system, who are the people, who are the least or most responsive, how the timing of certain lab results impacts the overall timeline of patient care, etc.

The next question was how to facilitate and simplify integration with newly acquired healthcare units in terms of systems and processes. In order to simplify integration with newly acquired healthcare, they use Web service technology to wrap individual systems as (REST) services and rely on a service bus and repository for their integration effort. From a process integration point of view, existing commercial BPMN-based process implementation tools (Table 6.1) can be used to integrate and compose individual services into a holistic healthcare process for HealthAlliance. In order to manage the process of mergers with new acquired units, process matching, merging, and version management methods introduced in *Chap. 3 are very essential.*

Through the above process analytics and process execution querying steps, the process optimization team at Acme working with HealthAlliance process owner's team updates their process model with the operational process information and insights. Also, they may apply a process mining and business process execution querying platform, described above, to build a knowledge graph over the healthcare process execution data from newly acquired businesses. The result of a similar analysis on those helps the Acme and HealthAlliance team to develop a process improvement strategy and a process integration strategy for onboarding new partners.

6.3 Discussion and Future Directions

6.3.1 Big Data Analytics for Process Data

The ability to harness the ever-increasing amounts of business-related data will enable us to understand what is happening in the world. In this context, "big data" is one of the biggest buzzwords around these days and it is going to impact BPs. In particular, generating huge metadata (e.g., versioning, provenance, security, and privacy) for imbuing the process data with additional semantics, the adoption of social media, the digitalization of business artifacts (e.g., files, documents, reports, and receipts), and using sensors (e.g., to track patients while out of hospital) will generate part of the big data.

In this context, understanding, analyzing, and ultimately improving big process data are the goals of enterprises today. These tasks are challenging as BPs in modern enterprises are implemented over several applications and Web services, and the information about process execution is scattered across several data sources. In particular, as large-scale processes are executed on cloud-based service-oriented environments, the process logs increasingly come to show all typical properties of the big data: wide physical distribution, diversity of formats, nonstandard data models, and independently managed and heterogeneous semantics.

Figure 6.1 illustrates a big picture for research directions in understanding big process data. Research directions in organizing process data include arranging data in a coherent form and systematizing its retrieval and processing. It is challenging as process data come from many sources and can be used for multiple purposes. These data inputs need to be organized and stored for computer processing. In this context, management, indexing, mining, and querying of large volumes of both structured and unstructured process-related data will be challenging. Research directions in process analytics include, but not limited to, performance, quality, and visualization which requires using crowdsourcing and storytelling techniques to understand the analytics results [27].

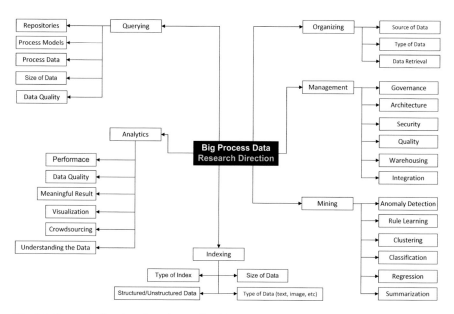

Fig. 6.1 Research directions in understanding big process data

Apart from the business-related data, various types of metadata may be collected on several systems and organizations, especially as processes require many different people (having lots of knowledge and experience) to collaborate to find the correct solution. In particular, in the process execution path, a huge amount of process-related metadata such as versioning (what are the various versions of an artifact, during its life cycle, and how they are related), provenance [58] (what manipulations were performed on the artifact to get it to this point), security (who has access to the artifact over time), and privacy (what actions were performed to protect or release artifact information over time) can be recorded. These metadata can be used for imbuing the process data with additional semantics and will manifest new challenges for process analysis.

In this context and to better understand business processes, there is a need to collect additional metadata such as the descriptions of where a business-related artifact came from, who accessed or updated it, and what its quality is [156]. The provenance [58] of the process-related data, what manipulations were performed on it to get it to this point, can also be recorded [58]. Similarly, the accuracy and lineage of the data can be captured [472]. Security and privacy introduce additional metadata, such as who has access to the data and to whom information has been released. Reliability and performance requirements are also potential metadata to be collected.

In order to capture such metadata, artifact-centric approaches have been proposed to model the changes and evolutions of business data or business-related entities [358]. Considering the metadata as a first-class citizen is of high importance, as business artifacts change over time and it is important to be able to query the evolution of artifacts over time. It is challenging as metadata annotations assigned to an artifact (or its versions) today may no longer be relevant to the future representation of that artifact: artifacts are very likely to have different states over time and the temporal annotations may or may not apply to these evolving states. Consequently, analyzing artifacts' metadata, e.g., evolving aspects of artifacts such as versioning and provenance, over time is important and will expose many hidden information among entities in process graphs. This information can be used to detect the actual processing behavior and, therefore, to improve business processes.

6.3.2 Analyzing Big Process Data Problem

Analyzing and understanding big process data offer important information for the organization's management. This information can be used to detect the actual processing behavior and, therefore, to improve the processes. However, understanding of business processes and analyzing BP execution data (e.g., logs containing events, interaction messages, and other process artifacts) are difficult as the information about process execution is scattered across several systems and data sources. Moreover, in many cases, there is no well-documented information on how this information is related to each other and to the overall business process of the enterprise [346].

The main barrier for understanding such processes is to identify the interactions among entities (e.g., process stakeholders and process artifacts) within BP execution data. In this context, most entities (structured or unstructured) in process logs are interconnected through rich semantic information, where entities and relationships among them can be modeled using graphs. Knowledge about business processes is often hidden in the relationships among entities in *process graphs*, i.e., BP execution data modeled using graphs. Understanding this hidden knowledge in terms of its scope and details is challenging especially as it is subjective, depending on the perspective of the process analyst. In the following, we discuss analyzing big process data challenges.

6.3.2.1 Organizing Big Process Data

In order to organize the big process data, the first step is gathering and integration of process execution data in a *process event log* from various, potentially heterogeneous, systems and services. The next step is providing techniques to identify entities (e.g., process stakeholders and process artifacts) and the interactions among them within such integrated process logs. In this context, most entities (structured or unstructured) in process logs are interconnected through rich semantic information, where entities and relationships among them can be modeled using graphs. Since graphs form a complex and expressive data type, there is a need for methods to organize and index the graph data. A new stream of work [117, 234, 250] used MapReduce [131] for processing huge amounts of unstructured data in a massively parallel way. Hadoop [470], the open-source implementation of MapReduce, provides a distributed file system (i.e., HDFS[9]) and a high-level language for data analysis, i.e., Pig.[10]

6.3.2.2 Indexing and Querying Big Process Data

In order to query process graphs, a graph query language is needed. A number of graph query languages [18] have been proposed in the literature. However, process-related data become increasingly large and cannot be analyzed using such systems. Parallel processing, e.g., using MapReduce framework, can be used to handle the big process data; however, many graph algorithms are very difficult to be parallelized [314]. Some works [235, 261] focused on query optimization techniques for the Apache Pig,[11] a high-level procedural language on top of Hadoop, and the MapReduce, to store and querying graphs. In addition to Pig, there are several high-level languages and environments for advanced MapReduce-like systems, including SCOPE [105], Sawzall [381], and Sphere [198]. As another challenge, graph queries are becoming extremely complected: queries against a complex ontology are often lengthy, regardless to the graph query language to be used. To answer these challenges, keyword search [18] and mining methods [296] have been used to simplify queries and to semiautomate the query formation. Moreover, data services have been proposed to simplify accessing, querying, and analyzing of information, e.g., related to processes.

6.3.2.3 Supporting Big Data Analytics over Process Execution Data

In today's knowledge-, service-, and cloud-based economy, businesses accumulate massive amounts of data from a variety of sources. In order to understand businesses, one may need to perform considerable analytics over large hybrid collections of heterogeneous and partially unstructured data that is captured related to the process execution. This data, usually modeled as graphs (i.e., process graphs), increasingly comes to show all the typical properties of *big data*: wide physical distribution, diversity of formats, nonstandard data models, and independently managed and heterogeneous semantics. OLAP of process graphs is challenging as the extension of existing OLAP techniques to the analysis of graphs is not straightforward. While traditional analytics solutions (relational DBs, data warehouses, and OLAP) do a great job in collecting data and providing answers on known

---

[9]http://hadoop.apache.org/

[10]http://pig.apache.org/

[11]http://pig.apache.org/

questions, key business insights remain hidden in the interactions among objects in process graphs. In particular, it will be hard to discover concept hierarchies for entities based on both data objects and their interactions in process graphs. Existing approaches [57, 162], in OLAP on graphs, took the first step by supporting multidimensional and multilevel queries on graphs; however, much work needs to be done to make OLAP networks a reality [215].

6.3.3 Crowdsourcing and Social BPM

In modern enterprises, collaboration and communication among business users fall outside of the BPM suite container. For example, email communication about a process, instant messaging to get a response to a process-related question, allowing business users to generate processes, and allowing frontline workers to update process knowledge (using new technologies such as process wikis) emphasize the social media's impact on business process management. For example, using social networks, users can interact with service providers to be informed, share experience, and express their opinion on the quality of a service.

In this context, business users can easily collaborate on doing jobs or share information through mass collaborative systems which are also called crowdsourcing systems [144]. Crowdsourcing has emerged as an effective way to perform tasks that are easy for humans but remain difficult for computers [26, 28]. For instance, Amazon Mechanical Turk (MTurk[12]) provides on-demand access to task forces for micro-tasks such as image recognition and language translation. Several organizations, including DARPA and various world health and relief agencies, are using platforms such as MTurk to crowdsource work through multiple channels, including SMS, email, Twitter, and the World Wide Web.

In this context, crowdsourcing can help organizations increase productivity by discovering and exploiting informal knowledge and relationships in order to improve activity execution. Crowdsourcing can also enable the social BPM to assign an activity to a broader set of performers or to find appropriate contributors for its execution. Moreover it will be possible to elicit opinions that contribute to taking a decision through acquiring feedback from a broader set of stakeholders and for process improvement.

The potential social BPM product space spans three families of solutions [169]: traditional BPM systems, enterprise social software, and emerging social BPM suites. Traditional BPM solutions do not support social interactions in business processes but could be integrated with social tools. Enterprise social software such as corporate wiki, blogs, and intranet portals augmented with user interaction capabilities enable social interactions but are not meant to enact articulated business

---

[12]https://www.mturk.com/

processes. Social BPM suites, such as IBM's Blueworks,[13] are promising but the actual social features are still under definition.

Social BPMs inevitably require advanced crowd-management capabilities in future social computing platforms. Possible research directions are discussed in [432] to extend: (a) process discovery and design to include interactive real-time involvement of business stakeholders, e.g., users, customers, and partners; (b) process development methodology and tools to support collaboration between business and IT roles; and (c) process design to provide real-time guidance for completing a particular activity based on real-time business analytics and social-network analysis using crowdsourcing [144] techniques.

6.3.4 Process Data Management in the Cloud

Management of process data is an important issue in rapidly changing organizations. For such organizations, it is important to increase processing/storage capacity or add software/hardware capabilities on the fly without investing in new infrastructure, training new personnel, or licensing new software. Cloud computing provides an infrastructure for these needs. The elasticity of the cloud enables the allocation of additional computational resources on the fly to handle the increased demand. In terms of storage, the cloud provides data availability, durability, and security through under-the-cover replication and encryption algorithms. In this context, process data management applications are potential candidates for deployment in the cloud.

In process data management, there is need to query different data stores to answer questions in business planning, problem solving, and decision support. In this context, the historical data along with data from multiple operational data sources are all typically involved in the analysis. Consequently, the scale of process data management systems is generally large and exceeds petabytes of process-related data. Existing available software solutions to perform the process data management includes MapReduce-like software and commercially available parallel databases.

MapReduce-like software include Apache Hadoop[14] solution, useful extensions [364], and Microsoft's Dryad/SCOPE stack [105]. All these software are designed to automate the parallelization of large-scale data analysis workloads. These software are designed with fault tolerance as a high priority through dividing a process data analysis job into many small tasks. Other characteristics [12] of MapReduce-like software include the ability to: run in a heterogeneous environment, operate on encrypted data, and interface with business intelligence products.

---

[13]https://www.blueworkslive.com

[14]http://hadoop.apache.org/

Another option for dealing with process data analysis in the cloud is the commercially available parallel databases, such as Teradata,[15] Netezza,[16] Vertica,[17] and Oracle Big Data.[18] These software hold a reasonable market share for on-premise large-scale data analysis. Compared to MapReduce-like software, parallel databases provide a better efficiency. It has been shown that [12] there is room for performance improvement in MapReduce-like software. To address the shortcomings, the Pig project at Yahoo [364] and the SCOPE project at Microsoft [105] aim to integrate declarative query constructs from the database community into MapReduce-like software. A possible research direction would be to combine the ease-of-use out-of-the-box advantages of MapReduce-like software with the efficiency and shared-work advantages that come with loading data and creating performance-enhancing data structures [12].

Another possible research direction will be to use graphs in cloud data management. Graphs are widely used for modeling complicated data in business processes. Modeling and analyzing big process data using graphs will generate graphs having millions and billions of nodes and edges. Using MapReduce to process such huge graphs will be challenging as MapReduce does not provide a direct support for iterative data analysis tasks: graph processing algorithms are iterative and need to traverse the graph in some way [45]. To address this challenge, Google proposed Pregel, a scalable platform for implementing graph algorithms. The Pregel system has been cloned by many open-source projects such as Apache Giraph,[19] Apache Hama,[20] and GoldenOrb.[21] Characterizing the performance of distributed graph computation platforms indicates that there is room for performance improvement: in domains such as graph processing, the efficiency and raw performance of MapReduce are a matter of debate.

---

[15]www.teradata.com/

[16]www.ibm.com/software/data/netezza/

[17]www.vertica.com/

[18]www.oracle.com/bigdata

[19]http://giraph.apache.org/

[20]http://hama.apache.org/

[21]https://github.com/jzachr/goldenorb

References

1. van der Aalst, W.M.P.: Process-aware information systems: Design, enactment, and analysis. In: Wiley Encyclopedia of Computer Science and Engineering. Wiley, New York (2008)
2. van der Aalst, W.M.P.: Process Mining: Discovery, Conformance and Enhancement of Business Processes. Springer, Heidelberg (2011)
3. van der Aalst, W.M.P.: Process mining: Overview and opportunities. ACM Trans. Manag. Inf. Syst. **3**(2), 7 (2012)
4. van der Aalst, W.M.P.: Service mining: Using process mining to discover, check, and improve service behavior. IEEE Trans. Serv. Comput. **99**(PrePrints), 1 (2012)
5. van der Aalst, W.M.P., Weijters, A.J.M.M.: Process mining: A research agenda. Comput. Ind. **53**(3), 231–244 (2004)
6. van der Aalst, W.M.P., van Dongen, B.F., Herbst, J., Maruster, L., Schimm, G., Weijters, A.J.M.M.: Workflow mining: A survey of issues and approaches. Data Knowl. Eng. **47**, 237–267 (2003)
7. van der Aalst, W.M.P., ter Hofstede, A.H.M., Weske, M.: Business process management: A survey. In: Business Process Management, pp. 1–12. Springer, Heidelberg (2003)
8. van der Aalst, W.M.P., Weijters, T., Maruster, L.: Workflow mining: Discovering process models from event logs. IEEE Trans. Knowl. Data Eng. **16**(9), 1128–1142 (2004)
9. van der Aalst, W.M.P., Weske, M., Grünbauer, D.: Case handling: A new paradigm for business process support. Data Knowl. Eng. **53**(2), 129–162 (2005)
10. van der Aalst, W.M.P., van Dongen, B.F., Günther, C.W., Rozinat, A., Verbeek, E., Weijters, T.: ProM: The process mining toolkit. In: Proceedings of the Business Process Management Demonstration Track (BPMDemos 2009), Ulm, 8 September 2009. CEUR Workshop Proceedings, vol. 489. Springer, Berlin (2009). CEUR-WS.org 2009
11. van der Aalst, W.M.P., Adriansyah, A., Medeiros, A.K.A., Arcieri, F., Baier, T., Blickle, T., Bose, R.P.J.C., van den Brand, P., Brandtjen, R., Buijs, J.C.A.M., Burattin, A., Carmona, J., Castellanos, M., Claes, J., Cook, J., Costantini, N., Curbera, F., Damiani, E., Leoni, M.D., Delias, P., van Dongen, B.F., Dumas, M., Dustdar, S., Fahland, D., Ferreira, D.R., Gaaloul, W., van Geffen, F., Goel, S., Günther, C.W., Guzzo, A., Harmon, P., ter Hofstede, A.H.M., Hoogland, J., Espen Ingvaldsen, J., Kato, K., Kuhn, R., Kumar, A., La Rosa, M., Maria Maggi, F., Malerba, D., Mans, R.S., Manuel, A., McCreesh, M., Mello, P., Mendling, J., Montali, M., Motahari-Nezhad, H.R., Muehlen, M.Z., Muñoz-Gama, J., Pontieri, L., Ribeiro, J., Rozinat, A., Pérez, H.S., Pérez, R.S., Sepúlveda, M., Sinur, J., Soffer, P., Song, M., Sperduti, A., Stilo, G., Stoel, C., Swenson, K.D., Talamo, M., Tan, W., Turner, C., Vanthienen, J., Varvaressos, G., Verbeek, E., Verdonk, M., Vigo, R., Wang, J., Weber, B., Weidlich, M., Weijters, T., Wen, L., Westergaard, M., Wynn, M.T.: Process mining manifesto. In: Business Process Management Workshops (1), pp. 169–194. Springer, Heidelberg (2011)

© Springer International Publishing Switzerland 2016

S.-M.-R. Beheshti et al., *Process Analytics*, DOI 10.1007/978-3-319-25037-3

12. Abadi, D.J.: Data management in the cloud: Limitations and opportunities. IEEE Data Eng. Bull. **32**(1), 3–12 (2009)
13. Abbaci, K., Lemos, F., HadjAli, A., Grigori, D., Lietard, L., Rocacher, D., Bouzeghoub, M.: Selecting and ranking business processes with preferences: An approach based on fuzzy sets. In: On the Move to Meaningful Internet Systems: OTM 2011 - Confederated International Conferences: CoopIS, DOA-SVI, and ODBASE 2011, Hersonissos, Crete, 17–21 October 2011, Proceedings, Part I, pp. 38–55 (2011)
14. Abecker, A., Bernardi, A., Ntioudis, S., Mentzas, G., Herterich, R., Houy, C., Müller, S., Legal, M.: The decor toolbox for workflow-embedded organizational memory access. Enterp. Inf. Syst. III **3**, 107 (2002)
15. Abelló, A., Romero, O.: On-line analytical processing. In: Encyclopedia of Database Systems, pp. 1949–1954. Springer, Heidelberg (2009)
16. Adi, A., Etzion, O.: Amit-the situation manager. VLDB J.—Int. J. Very Large Data Bases **13**(2), 177–203 (2004)
17. Adriansyah, A., van Dongen, B.F., van der Aalst, W.M.P.: Conformance checking using cost-based fitness analysis. In: Proceedings of the 15th IEEE International Enterprise Distributed Object Computing Conference (EDOC 2011), Helsinki, 29 August–2 September 2011, pp. 55–64. IEEE Computer Society, Los Alamitos (2011) [ISBN 978-1-4577-0362-1]
18. Aggarwal, C.C., Wang, H.: Managing and Mining Graph Data. Springer, Heidelberg (2010)
19. Agostini, A., De Michelis, G.: Improving flexibility of workflow management systems. In: Business Process Management, pp. 218–234. Springer, Heidelberg (2000)
20. Agrawal, A.: Semantics of business process vocabulary and process rules. In: Proceedings of the 4th India Software Engineering Conference, pp. 61–68. ACM, New York (2011)
21. Agrawal, R., Gunopulos, D., Leymann, F.: Mining process models from workflow logs. In: Proceedings of Advances in Database Technology - EDBT'98. 6th International Conference on Extending Database Technology, Valencia, 23–27 March 1998. Lecture Notes in Computer Science, vol. 1377, pp. 469–483. Springer, New York (1998) [ISBN 3-540-64264-1]
22. Aho, A.V., Hopcroft, J.E., Ullman, J.D.: Data Structures and Algorithms. Computer Science and Information Processing. Addison-Wesley, Reading (1983)
23. Akal, F., Bim, K., Schek, H.J.: OLAP query evaluation in a database cluster: A performance study on intra-query parallelism. In: Proceedings of the 6th East European Conference on Advances in Databases and Information Systems (ADBIS 2002), Bratislava, 8–11 September 2002. Lecture Notes in Computer Science, vol. 2435, pp. 218–231. Springer, Berlin (2002) [ISBN 3-540-44138-7]
24. Albek, E., Bax, E., Billock, G., Chandy, K.M., Swett, I.: An event processing language (epl) for building sense and respond applications. In: Proceedings of the 19th IEEE International Parallel and Distributed Processing Symposium, 2005, pp. 136b–136b. IEEE, New York (2005)
25. Ali, S., Torabi, T., Soh, B.: Rule component specification for business process deployment. In: 18th International Workshop on Database and Expert Systems Applications, 2007 (DEXA'07), pp. 595–599. IEEE, New York (2007)
26. Allahbakhsh, M., Ignjatovic, A., Benatallah, B., Beheshti, S.-M.-R., Bertino, E., Foo, N.: Reputation management in crowdsourcing systems. In: 8th International Conference on Collaborative Computing: Networking, Applications and Worksharing, CollaborateCom 2012, Pittsburgh, 14–17 October 2012, pp. 664–671 (2012)
27. Allahbakhsh, M., Ignjatovic, A., Benatallah, B., Beheshti, S.-M.-R., Bertino, E., Foo, N.: Collusion detection in online rating systems. In: Proceedings of the 15th Asia-Pacific Web Conference on Web Technologies and Applications, APWeb 2013, Sydney, 4–6 April 2013, pp. 196–207 (2013)
28. Allahbakhsh, M., Ignjatovic, A., Benatallah, B., Beheshti, S.-M.-R., Foo, N., Bertino, E.: Representation and querying of unfair evaluations in social rating systems. Comput. Secur. **41**, 68–88 (2014)
29. Alonso, G., Casati, F., Kuno, H.A., Machiraju, V.: Web services - concepts, architectures and applications. In: Data-Centric Systems and Applications. Springer, Heidelberg (2004)

30. Alves, A., Arkin, A., Askary, I.S., Barreto, I.C., Bloch, B., Curbera, F., Ford, M., Goland, Y., Guiar, A., Kartha, N., Liu, C.K., Khalaf, R., Kig, D., Marin, M., Mehta, V., Thatte, S., van der Rijn, D., Yendluri, P., Yiu, A.: WSBPEL:Web Services Business Process Execution Language, Version 2.0. http://docs.oasis-open.org/wsbpel/2.0/ (2006)
31. Ammon, R.V., Silberbauer, C., Wolff, C.: Domain specific reference models for event patterns—for faster developing of business activity monitoring applications. In: VIP Symposia on Internet Related Research with Elements of M+ I+ T+, Lake Bled, vol. 16 (2007)
32. Angell, R.C., Freund, G.E., Willett, P.: Automatic spelling correction using a trigram similarity measure. Inf. Process. Manag. **19**(4), 255–261 (1983)
33. Anicic, D., Fodor, P., Rudolph, S., Stojanovic, N.: EP-SPARQL: A unified language for event processing and stream reasoning. In: Proceedings of the 20th International Conference on World Wide Web (WWW 2011), Hyderabad, 28 March–1 April 2011. ACM, New York (2011) [ISBN 978-1-4503-0632-4]
34. Arasu, A., Chaudhuri, S., Kaushik, R.: Transformation-based framework for record matching. In: Proceedings of the 24th International Conference on Data Engineering, ICDE 2008, 7–12 April 2008, Cancún, pp. 40–49 (2008)
35. Ardissono, L., Furnari, R., Goy, A., Petrone, G., Segnan, M.: Fault tolerant web service orchestration by means of diagnosis. In: Third European Workshop on Software Architecture (EWSA 2006), Nantes, 4–5 September 2006, Revised Selected Papers. Lecture Notes in Computer Science, vol. 4344, pp. 2–16. Springer, Berlin (2006) [ISBN 3-540-69271-1]
36. Awad, A.: BPMN-Q: A Language to Query Business Processes. In: Enterprise Modelling and Information Systems Architectures - Concepts and Applications. Proceedings of the 2nd International Workshop on Enterprise Modelling and Information Systems Architectures (EMISA'07), St. Goar, 8–9 October 2007, pp. 115–128 [LNI P-119, GI 2007, ISBN 978-3-88579-213-0]
37. Awad, A., Polyvyanyy, A., Weske, M.: Semantic querying of business process models. In: 12th International IEEE Enterprise Distributed Object Computing Conference (ECOC 2008), Munich, 15–19 September 2008, pp. 85–94
38. Azvine, B., Nauck, D., Ho, C.: Intelligent business analytics - a tool to build decision-support systems for ebusinesses. BT Technol. J. **21**(4), 65–71 (2003)
39. Babcock, B., Babu, S., Datar, M., Motwani, R., Widom, J.: Models and issues in data stream systems. In: Proceedings of the Twenty-First ACM SIGMOD-SIGACT-SIGART Symposium on Principles of Database Systems, pp. 1–16. ACM, New York (2002)
40. Báez, M., Parra, C., Casati, F., Marchese, M., Daniel, F., di Meo, K., Zobele, S., Menapace, C., Valeri, B.: Gelee: Cooperative lifecycle management for (composite) artifacts. In: Service-Oriented Computing, pp. 645–646. Springer, Heidelberg (2009)
41. Bahiense, L., Manic, G., Piva, B., de Souza, C.C.: The maximum common edge subgraph problem: A polyhedral investigation. Discret. Appl. Math. **160**(18), 2523–2541 (2012)
42. Baier, T., Mendling, J.: Bridging abstraction layers in process mining by automated matching of events and activities. In: Proceedings of the 11th International Conference on Business Process Management, BPM 2013, Beijing, 26–30 August 2013, pp. 17–32 (2013)
43. Balmin, A., Papadimitriou, T., Papakonstantinou, Y.: Hypothetical queries in an OLAP environment. In: Proceedings of the 16th International Conference on Data Engineering, San Diego, 28 February–3 March 2000, pp. 220–231. IEEE Computer Society, Los Alamitos (2000) [ISBN 0-7695-0506-6]
44. Barbon, F., Traverso, P., Pistore, M., Trainotti, M.: Run-time monitoring of instances and classes of web service compositions. In: IEEE International Conference on Web Services (ICWS 2006), Chicago, 18–22 September 2006, pp. 63–71
45. Barnawi, A., Batarfi, O., Beheshti, S.-M.-R., Elshawi, R., Nouri, R., Sakr, S.: On characterizing the performance of distributed graph computation platforms. In: TPC Technology Conference (TPCTC 2014), Hangzhou. Lecture Notes in Computer Science (2014)
46. Barukh, M.C., Benatallah, B.: Servicebase: A programming knowledge-base for service oriented development. In: Database Systems for Advanced Applications, pp. 123–138. Springer, Heidelberg (2013)

47. Barukh, M.C., Benatallah, B.: A toolkit for simplified web-services programming. In: Web Information Systems Engineering–WISE 2013, pp. 515–518. Springer, Heidelberg (2013)
48. Barukh, M.C., Benatallah, B.: Processbase: A hybrid process management platform. In: Service-Oriented Computing, pp. 16–31. Springer, Heidelberg (2014)
49. Beardsley, S.C., Johnson, B.C., Manyika, J.M.: Competitive advantage from better interactions. McKinsey Q. **2**, 52 (2006)
50. Becker, M., Laue, R.: A comparative survey of business process similarity measures. Comput. Ind. **63**(2), 148–167 (2012)
51. Beeri, C., Eyal, A., Kamenkovich, S., Milo, T.: Querying business processes with BP-QL. In: Proceedings of the 31st International Conference on Very Large Data Bases, Trondheim, 30 August–2 September 2005, pp. 1255–1258
52. Beeri, C., Eyal, A., Milo, T., Pilberg, A.: Monitoring business processes with queries. In: Proceedings of the 33rd International Conference on Very Large Data Bases, University of Vienna, Vienna, 23–27 September 2007, pp. 603–614. ACM, New York (2007) [ISBN 978-1-59593-649-3]
53. Beeri, C., Eyal, A., Milo, T., Pilberg, A.: BP-Mon: Query-based monitoring of BPEL business processes. SIGMOD Rec. **37**(1), 21–24 (2008)
54. Beheshti, S.-M.-R.: Organizing, Querying, and Analyzing Ad-hoc Processes' Data. Ph.D. thesis, University of New South Wales, Sydney (2012)
55. Beheshti, S.-M.-R., Moshkenani, M.S.: Development of grid resource discovery service based on semantic information. In: Proceedings of the 2007 Spring Simulation Multiconference, SpringSim 2007, Norfolk, 25–29 March 2007, vol. 1, pp. 141–148 (2007)
56. Beheshti, S.-M.-R., Benatallah, B., Nezhad, H.R.M., Sakr, S.: A query language for analyzing business processes execution. In: Proceedings of the 9th International Conference on Business Process Management, BPM 2011, Clermont-Ferrand, 30 August–2 September 2011, pp. 281–297 (2011)
57. Beheshti, S.-M.-R., Benatallah, B., Nezhad, H.R.M., Allahbakhsh, M.: A framework and a language for on-line analytical processing on graphs. In: Proceedings of the 13th International Conference on Web Information Systems Engineering, WISE 2012, Paphos, 28–30 November 2012, pp. 213–227 (2012)
58. Beheshti, S.-M.-R., Nezhad, H.R.M., Benatallah, B.: Temporal provenance model (tpm): Model and query language. CoRR, abs/1211.5009 (2012)
59. Beheshti, S.-M.-R., Sakr, S., Benatallah, B., Nezhad, H.R.M.: Extending SPARQL to support entity grouping and path queries. CoRR (2012)
60. Beheshti, S.-M.-R., Benatallah, B., Nezhad, H.R.M.: Enabling the analysis of cross-cutting aspects in ad-hoc processes. In: Proceedings of the 25th International Conference on Advanced Information Systems Engineering, CAiSE 2013, Valencia, 17–21 June 2013, pp. 51–67 (2013)
61. Beheshti, S.-M.-R., Benatallah, B., Motahari-Nezhad, H.: Scalable graph-based olap analytics over process execution data. Distrib. Parallel Databases 1–45 (2015). doi:10.1007/s10619-014-7171-9
62. Beheshti, S.-M.-R., Venugopal, S., Ryu, S.H., Benatallah, B., Wang, W.: Big Data and Cross-Document Coreference Resolution: Current State and Future Opportunities (2013). CoRR abs/1311.3987
63. Behrends, E., Fritzen, O., May, W., Schubert, D.: An eca engine for deploying heterogeneous component languages in the semantic web. In: Current Trends in Database Technology–EDBT 2006, pp. 887–898. Springer, Heidelberg (2006)
64. Bellotti, V., Ducheneaut, N., Howard, M., Smith, I.: Taking email to task: The design and evaluation of a task management centered email tool. In: Proceedings of the 2003 Conference on Human Factors in Computing Systems (CHI 2003), Ft. Lauderdale, 5–10 April 2003, pp. 345–352. ACM, New York (2003) [ISBN 1-58113-630-7]
65. Benatallah, B., Hacid, M.S., Rey, C., Toumani, F.: Semantic reasoning for web services discovery. In: Proceedings of the Twelfth International World Wide Web Conference (WWW 2003), Budapest, 20–24 May 2003. ACM, New York (2003) [ISBN 1-58113-680-3]

66. Benatallah, B., Casati, F., Grigori, D., Nezhad, H.R.M., Toumani, F.: Developing adapters for web services integration. In: Proceedings of 17th International Conference on Advanced Information Systems Engineering (CAiSE 2005), Porto, 13–17 June 2005, pp. 415–429
67. Benatallah, B., Dumas, M., Sheng, Q.Z.: Facilitating the rapid development and scalable orchestration of composite web services. Distrib. Parallel Databases **17**(1), 5–37 (2005)
68. Benatallah, B., Casati, F., Toumani, F.: Representing, analysing and managing web service protocols. Data Knowl. Eng. **58**, 327–357 (2006)
69. Berardi, D., De Rosa, F., De Santis, L., Mecella, M.: Finite state automata as conceptual model for e-services. J. Integr. Des. Process. Sci. Arch. **8**, 105–121 (2004)
70. Bergamaschi, S., Castano, S., Vincini, M.: Semantic integration of semistructured and structured data sources. SIGMOD Rec. **28**(1), 54–59 (1999)
71. Bergamaschi, S., Castano, S., Vincini, M., Beneventano, D.: Semantic integration of heterogeneous information sources. Data Knowl. Eng. **36**(3), 215–249 (2001)
72. Berlin, J., Motro, A.: Database schema matching using machine learning with feature selection. In: Proceedings of 14th International Conference on Advanced Information Systems Engineering (CAiSE 2002), Toronto, 27–31 May 2002, pp. 452–466
73. Bernstein, A.: How can cooperative work tools support dynamic group process? bridging the specificity frontier. In: Proceeding on the ACM 2000 Conference on Computer Supported Cooperative Work (CSCW 2000), Philadelphia, 2–6 December 2000, pp. 279–288. ACM, New York (2000) [ISBN 1-58113-222-0]
74. Bernstein, P.A., Melnik, S., Petropoulos, M., Quix, C.: Industrial-strength schema matching. SIGMOD Rec. **33**(4), 38–43 (2004)
75. Berstel, B., Bonnard, P., Bry, F., Eckert, M., Pătrânjan, P.-L.: Reactive rules on the web. In: Reasoning Web, pp. 183–239. Springer, Heidelberg (2007)
76. Beyer, K.S., Ramakrishnan, R.: Bottom-up computation of sparse and iceberg CUBEs. In: SIGMOD 1999, Proceedings ACM SIGMOD International Conference on Management of Data, 1–3 June 1999, Philadelphia, pp. 359–370. ACM, New York (1999)
77. Bhalotia, G., Hulgeri, A., Nakhe, C., Chakrabarti, S., Sudarshan, S.: Keyword searching and browsing in databases using BANKS. In: Proceedings of the 18th International Conference on Data Engineering, San Jose, 26 February–1 March 2002, pp. 431–440
78. Bhattacharya, K., Guttman, R., Lyman, K., Heath III, F.F., Kumaran, S., Nandi, P., Wu, F., Athma, P., Freiberg, C., Johannsen, L., et al.: A model-driven approach to industrializing discovery processes in pharmaceutical research. IBM Syst. J. **44**(1), 145–162 (2005)
79. Bhattacharya, K., Caswell, N.S., Kumaran, S., Nigam, A., Wu, F.Y.: Artifact-centered operational modeling: Lessons from customer engagements. IBM Syst. J. **46**(4), 703–721 (2007)
80. Bhattacharya, K., Gerede, C., Hull, R., Liu, R., Su, J.: Towards formal analysis of artifact-centric business process models. In: Business Process Management (BPM'07), pp. 288–304. Springer, Heidelberg (2007)
81. Bhattacharya, K., Hull, R., Su, J.: A data-centric design methodology for business processes. In: Handbook of Research on Business Process Modeling, Chapter 23, pp. 503–531. Information Science Reference, Hershey (2009)
82. Bifet, A., Holmes, G., Kirkby, R., Pfahringer, B.: Moa: Massive online analysis. J. Mach. Learn. Res. **11**, 1601–1604 (2010)
83. Bilenko, M., Mooney, R.J., Cohen, W.W., Ravikumar, P.D., Fienberg, S.E.: Adaptive name matching in information integration. IEEE Intell. Syst. **18**(5), 16–23 (2003)
84. Bizer, C., Heath, T., Berners-Lee, T.: Linked data - the story so far. Int. J. Semant. Web Inf. Syst. **5**(3), 1–22 (2009)
85. Blaze advisor: http://www.fico.com/en/products/fico-blaze-advisor-business-rules-management-system/. Accessed July 2015
86. Böhm, A., Marth, E., Kanne, C.-C.: The demaq system: Declarative development of distributed applications. In: Proceedings of the 2008 ACM SIGMOD International Conference on Management of Data, pp. 1311–1314. ACM, New York (2008)

87. Böhringer, M.: Emergent case management for ad-hoc processes: A solution based on microblogging and activity streams. In: Business Process Management Workshops, pp. 384–395. Springer, Heidelberg (2011)

88. Bonifati, A., Paraboschi, S.: Active xquery. In: Web Dynamics, pp. 249–274. Springer, Heidelberg (2004)

89. Bose, R.P.J.C., Verbeek, H.M.W., van der Aalst, W.M.P.: Discovering hierarchical process models using ProM. In: Proceedings of the CAiSE Forum 2011, London, 22–24 June 2011, pp. 33–40

90. Bouman, W., de Bruin, B., Hoogenboom, T., Huizing, A., Jansen, R., Schoondorp, M.: The realm of sociality: Notes on the design of social software (2008). In: Proceedings of the International Conference on Information Systems (ICIS 2007), Montreal, 9–12 December 2007

91. Brambilla, M., Fraternali, P., Vaca, C.: BPMN and design patterns for engineering social BPM solutions. In: Business Process Management Workshops. Lecture Notes in Business Information Processing, vol. 99, pp. 219–230. Springer, Berlin (2012)

92. Brandl, H.M., Guschakowski, D.: Complex Event Processing in the context of Business Activity Monitoring. An evaluation of different approaches and tools taking the example of the Next Generation easyCredit. Ph.D. thesis, Diploma thesis, Preworkshop DEBS 2007 (2007)

93. Buse, R.P.L., Zimmermann, T.: Information needs for software development analytics. In: 34th International Conference on Software Engineering (ICSE 2012), Zurich, 2–9 June 2012, pp. 987–996

94. Cai, Y., Dong, X.L., Halevy, A.Y., Liu, J.M., Madhavan, J.: Personal information management with SEMEX. In: Proceedings of the ACM SIGMOD International Conference on Management of Data, Baltimore, 14–16 June 2005, pp. 921–923

95. Calders, T., Günther, C.W., Pechenizkiy, M., Rozinat, A.: Using minimum description length for process mining. In: Proceedings of the 2009 ACM Symposium on Applied Computing (SAC), Honolulu, 9–12 March 2009, pp. 1451–1455. ACM, New York (2009) [ISBN 978-1-60558-166-8]

96. Cantara, M.: User survey analysis: Soa, web services and web 2.0 user adoption trends and recommendations for software vendors, North America and Europe, 2005–2006 (2007)

97. Cardoso, J., Sheth, A.: Semantic e-workflow composition. J. Intell. Inf. Syst. **21**, 191–225 (2003)

98. Carey, M.J.: SOA what? IEEE Comput. **41**(3), 92–94 (2008)

99. Carey, M.J., Onose, N., Petropoulos, M.: Data services. Commun. ACM **55**(6), 86–97 (2012)

100. Casati, F., Shan, M.C.: Semantic analysis of business process executions. In: Advances in Database Technology - EDBT 2002, Proceedings of 8th International Conference on Extending Database Technology, Prague, 25–27 March 2002. Lecture Notes in Computer Science, vol. 2287, pp. 287–296. Springer, Berlin (2002) [ISBN 3-540-43324-4]

101. Casati, F., Ceri, S., Pernici, B., Pozzi, G.: Workflow evolution. Data Knowl. Eng. **24**(3), 211–238 (1998)

102. Casati, F., Castellanos, M., Dayal, U., Salazar, N.: A generic solution for warehousing business process data. In: Proceedings of the 33rd International Conference on Very Large Data Bases, University of Vienna, Vienna, 23–27 September 2007, pp. 1128–1137. ACM, New York (2007) [ISBN 978-1-59593-649-3]

103. Cayoglu, U., Dijkman, R.M., Dumas, M., Fettke, P., García-Bañuelos, L., Hake, P., Klinkmüller, C., Leopold, H., Ludwig, A., Loos, P., Mendling, J., Oberweis, A., Schoknecht, A., Sheetrit, E., Thaler, T., Ullrich, M., Weber, I., Weidlich, M.: Report: The process model matching contest 2013. In: Business Process Management Workshops - BPM 2013 International Workshops, Beijing, 26 August 2013, Revised Papers, pp. 442–463 (2013)

104. Cearley, D.: Gartner identifies the top 10 strategic technology trends for 2013. In: Gartner Symposium/ITxpo, Orlando, pp. 6–10 (2013)

105. Chaiken, R., Jenkins, B., Larson, P., Ramsey, B., Shakib, D., Weaver, S., Zhou, J.: SCOPE: Easy and efficient parallel processing of massive data sets. PVLDB **1**(2), 1265–1276 (2008)

106. Chang, F., Dean, J., Ghemawat, S., Hsieh, W.C., Wallach, D.A., Burrows, M., Chandra, T., Fikes, A., Gruber, R.E.: Bigtable: A distributed storage system for structured data. ACM Trans. Comput. Syst. **26**(2), 4 (2008)

107. Chaudhuri, S., Dayal, U.: An overview of data warehousing and OLAP technology. SIGMOD Rec. **26**(1), 65–74 (1997)

108. Chaudhuri, S., Dayal, U., Narasayya, V.: An overview of business intelligence technology. Commun. ACM **54**(8), 88–98 (2011)

109. Chebotko, A., Lu, S., Fotouhi, F.: Semantics preserving SPARQL-to-SQL translation. Data Knowl. Eng. **68**(10), 973–1000 (2009)

110. Chebotko, A., Lu, S., Fei, X., Fotouhi, F.: RDFProv: A relational RDF store for querying and managing scientific workflow provenance. Data Knowl. Eng. **69**(8), 836–865 (2010)

111. Chen, C., Yan, X., Zhu, F., Han, J., Yu, P.S.: Graph OLAP: Towards online analytical processing on graphs. In: Proceedings of the 8th IEEE International Conference on Data Mining (ICDM 2008), Pisa, 15–19 December 2008, pp. 103–112

112. Cheney, J., Chiticariu, L., Tan, W.C.: Provenance in databases: Why, how, and where. Found. Trends Databases **1**, 379–474 (2009)

113. Chisholm, M.: How to Build a Business Rules Engine: Extending Application Functionality Through Metadata Engineering. Morgan Kaufmann, Los Altos (2004)

114. Choi, I., Kim, K., Jang, M.: An XML-based process repository and process query language for integrated process management. Knowl. Process Manag. **14**(4), 303–316 (2007)

115. Christen, P.: A comparison of personal name matching: Techniques and practical issues. In: Workshops Proceedings of the 6th IEEE International Conference on Data Mining (ICDM 2006), Hong Kong, 18–22 December 2006, pp. 290–294

116. Coalition, W.M.: Terminology and Glossary. Document Number WFMCTC-1011. www.wfmc.org/standards/docs/TC-1011termglossaryv3.pdf (1999)

117. Cohen, J.: Graph twiddling in a MapReduce world. Comput. Sci. Eng. **11**(4), 29–41 (2009)

118. Cohn, D., Hull, R.: Business artifacts: A data-centric approach to modeling business operations and processes. IEEE Data Eng. Bull. **32**(3), 3–9 (2009)

119. Cohn, D., Dhoolia, P., Heath Iii, F., Pinel, F., Vergo, J.: Siena: From powerpoint to web app in 5 minutes. In: Service-Oriented Computing–ICSOC 2008, pp. 722–723. Springer, Berlin (2008)

120. Colucci, S., Di Noia, T., Di Sciascio, E., Donini, F.M., Mongiello, M.: Concept abduction and contraction for semantic-based discovery of matches and negotiation spaces in an e-marketplace. Electron. Commer. Res. Appl. **4**(4), 345–361 (2005)

121. Consens, M.P., Mendelzon, A.O.: The G+/GraphLog Visual Query System. In: Special Interest Group on Management of Data Conference, Atlantic City, p. 388 (1990)

122. Cook, J.E., Wolf, A.L.: Discovering models of software processes from event-based data. ACM Trans. Softw. Eng. Methodol. **7**(3), 215–249 (1998)

123. Cooper, B.F., Ramakrishnan, R., Srivastava, U., Silberstein, A., Bohannon, P., Jacobsen, H.A., Puz, N., Weaver, D., Yerneni, R.: PNUTS: Yahoo!'s hosted data serving platform. PVLDB **1**(2), 1277–1288 (2008)

124. Cugola, G., Margara, A.: Processing flows of information: From data stream to complex event processing. ACM Comput. Surv. (CSUR) **44**(3), 15 (2012)

125. Curbera, F., Khalaf, R., Nagy, W., Weerawarana, S.: Implementing BPEL4WS: The architecture of a BPEL4WS implementation. Concurrency Comput.: Pract. Exp. **18**(10), 1219–1228 (2006)

126. Curran, T.A., Ladd, A.: SAP R/3 Business Blueprint: Understanding Enterprise Supply Chain Management, 2nd edn. Prentice Hall, Englewood Cliffs (1999)

127. Datta, A.: Automating the discovery of AS-IS business process models: probabilistic and algorithmic approaches. Inf. Syst. Res. **9**(3), 275–301 (1998)

128. Davenport, T.H.: Thinking for a Living: How to Get Better Performances and Results from Knowledge Workers. Harvard Business Press, Boston (2005)

129. De Giacomo, G., De Masellis, R., Grasso, M., Maggi, F.M., Montali, M.: Monitoring business metaconstraints based on LTL and LDL for finite traces. In: Proceedings of the

12th International Conference on Business Process Management, BPM 2014, Haifa, 7–11 September 2014, pp. 1–17 (2014)

130. de Man, H.: Case management: A review of modelling approaches. BPTrends: White paper (January 2009). http://www.ww.bptrends.com/publicationfiles/01-09-ART-%20Case %20Management-1-DeMan.%20doc--final.pdf

131. Dean, J., Ghemawat, S.: MapReduce: Simplified data processing on large clusters. Commun. ACM **51**(1), 107–113 (2008)

132. DeCandia, G., Hastorun, D., Jampani, M., Kakulapati, G., Lakshman, A., Pilchin, A., Sivasubramanian, S., Vosshall, P., Vogels, W.: Dynamo: Amazon's highly available key-value store. In: Proceedings of the 21st ACM Symposium on Operating Systems Principles 2007 (SOSP 2007), Stevenson, Washington, 14–17 October 2007, pp. 205–220. ACM, New York (2007) [ISBN 978-1-59593-591-5]

133. Delbru, R., Campinas, S., Tummarello, G.: Searching web data: An entity retrieval and high-performance indexing model. J. Web Semant. **10**, 33–58 (2012)

134. Delbru, R., Toupikov, N., Catasta, M., Tummarello, G.: A node indexing scheme for Web entity retrieval. In: Proceedings of the 7th Extended Semantic Web Conference (ESWC 2010), Part II, Heraklion, 30 May–3 June 2010, pp. 240–256

135. Delp, M., Böhm, K., Engelbach, W.: Pre-built information space: Some observations on the challenges of process-oriented knowledge management. In: 14th International Conference on Cognitive Computing and Data-Driven Business, Graz, September 2014

136. Dhamankar, R., Lee, Y., Doan, A., Halevy, A.Y., Domingos, P.: imap: Discovering complex mappings between database schemas. In: Proceedings of the ACM SIGMOD International Conference on Management of Data, Paris, 13–18 June 2004, pp. 383–394 (2004)

137. Dijkman, R.M., Dumas, M., García-Bañuelos, L.: Graph matching algorithms for business process model similarity search. In: 7th International Conference on Business Process Management (BPM 2009), Ulm, 8–10 September. Lecture Notes in Computer Science, vol. 5701, pp. 48–63. Springer, Berlin (2009)

138. Dijkman, R.M., Dumas, M., García-Bañuelos, L., Käärik, R.: Aligning business process models. In: Proceedings of the 13th IEEE International Enterprise Distributed Object Computing Conference, 1–4 September 2009, Auckland, pp. 45–53. IEEE Computer Society, Los Alamitos (2009)

139. Dijkman, R.M., La Rosa, M., Reijers, H.A.: Managing large collections of business process models - current techniques and challenges. Comput. Ind. **63**(2), 91–97 (2012)

140. Dimitrov, M., Simov, A., Stein, S., Konstantinov, M.: A BPMO based semantic business process modelling environment. In: Proceedings of the Workshop on Semantic Business Process and Product Lifecycle Management (SBPM 2007), Held in Conjunction with the 3rd European Semantic Web Conference (ESWC 2007), Innsbruck, 7 June 2007

141. Do, H.H., Rahm, E.: COMA - a system for flexible combination of schema matching approaches. In: Proceedings of 28th International Conference on Very Large Data Bases (VLDB 2002), Hong Kong, 20–23 August 2002, pp. 610–621 (2002)

142. Doan, A., Domingos, P., Halevy, A.Y.: Reconciling schemas of disparate data sources: A machine-learning approach. In: Special Interest Group on Management of Data Conference, Santa Barbara, pp. 509–520 (2001)

143. Doan, A., Madhavan, J., Domingos, P., Halevy, A.: Learning to map between ontologies on the semantic web. In: Proceedings of the Eleventh International World Wide Web Conference (WWW 2002), Honolulu, 7–11 May 2002, pp. 662–673

144. Doan, A., Ramakrishnan, R., Halevy, A.Y.: Crowdsourcing systems on the World-Wide Web. Commun. ACM **54**(4), 86–96 (2011)

145. Dong, X., Halevy, A.Y.: Indexing dataspaces. In: Proceedings of the ACM SIGMOD International Conference on Management of Data, Beijing, 12–14 June 2007, pp. 43–54

146. Dong, X., Halevy, A.Y., Madhavan, J., Nemes, E., Zhang, J.: Similarity search for web services. In: (e)Proceedings of the Thirtieth International Conference on Very Large Data Bases (VLDB 2004), Toronto, 31 August–3 September 2004, pp. 372–383 (2004)

147. Dong, Z., Wen, L., Huang, H., Wang, J.: CFS: a behavioral similarity algorithm for process models based on complete firing sequences. In: On the Move to Meaningful Internet Systems: OTM 2014 Conferences - Confederated International Conferences: CoopIS, and ODBASE 2014, Amantea, 27–31 October 2014, Proceedings. Lecture Notes in Computer Science, vol. 8841, pp. 202–219 (2014)
148. van Dongen, B.F., Mendling, J., van der Aalst, W.M.P.: Structural patterns for soundness of business process models. In: Tenth IEEE International Enterprise Distributed Object Computing Conference (EDOC 2006), Hong Kong, 16–20 October 2006, pp. 116–128. IEEE Computer Society, Los Alamitos (2006) [ISBN 0-7695-2558-X]
149. Dorn, C., Burkhart, T., Werth, D., Dustdar, S.: Self-adjusting recommendations for people-driven ad-hoc processes. In: Proceedings of 8th International Conference on Business Process Management (BPM 2010), Hoboken, 13–16 September 2010, pp. 327–342. Lecture Notes in Computer Science, vol. 6336. Springer, Berlin (2010) [ISBN 978-3-642-15617-5]
150. Dorn, C., Mar, C.A., Mehandjiev, N., Dustdar, S.: Self-learning predictor aggregation for the evolution of people-driven ad-hoc processes. In: Proceedings of 9th International Conference on Business Process Management (BPM 2011), Clermont-Ferrand, 30 August–2 September 2011, pp. 215–230. Lecture Notes in Computer Science, vol. 6896. Springer, Berlin (2011) [ISBN 978-3-642-23058-5]
151. Dries, A., Nijssen, S., De Raedt, L.: A query language for analyzing networks. In: Conference on Information and Knowledge Management (CIKM'09), pp. 485–494. ACM, New York (2009)
152. Dumais, S., Cutrell, E., Cadiz, J., Jancke, G., Sarin, R., Robbins, D.C.: Stuff i've seen: A system for personal information retrieval and re-use. In: Proceedings of the 26th Annual International ACM SIGIR Conference on Research and Development in Information Retrieval, SIGIR '03, pp. 72–79. ACM, New York (2003)
153. Dumas, M., Spork, M., Wang, K.: Adapt or perish: Algebra and visual notation for service interface adaptation. In: Proceedings of the 4th International Conference on Business Process Management, BPM 2006, Vienna, 5–7 September 2006, pp. 65–80 (2006)
154. Dumas, M., García-Bañuelos, L., Dijkman, R.M.: Similarity search of business process models. IEEE Data Eng. Bull. **32**(3), 23–28 (2009)
155. Dustdar, S., Hoffmann, T., van der Aalst, W.M.P.: Mining of ad-hoc business processes with teamlog. Data Knowl. Eng. **55**(2), 129–158 (2005)
156. Dyreson, C.E.: Aspect-oriented relational algebra. In: Proceedings of 14th International Conference on Extending Database Technology (EDBT 2011), Uppsala, 21–24 March 2011, pp. 377–388. ACM, New York (2011) [ISBN 978-1-4503-0528-0]
157. Ehrig, M., Koschmider, A., Oberweis, A.: Measuring similarity between semantic business process models. In: Asia-Pacific Conference on Conceptual Modelling, Ballarat, pp. 71–80 (2007)
158. Ellis, C.A., Keddara, K.: A workflow change is a workflow. In: Business Process Management, pp. 201–217. Springer, Berlin (2000)
159. Embley, D.W., Jackman, D., Xu, L.: Multifaceted exploitation of metadata for attribute match discovery in information integration. In: Proceedings of the International Workshop on Information Integration on the Web, Rio de Janeiro, 9–11 April 2001, pp. 110–117 [ISBN 85-901839-1-2]
160. Eric Prud'hommeaux, A.S.: Sparql query language for rdf, version 1.1. In: Standards Proposal by W3C and Hewlett-Packard Laboratories. W3C (2008). https://www.w3.org/TR/rdf-sparql-query/
161. Eshuis, R., Grefen, P.W. P.J.: Structural matching of BPEL processes. In: Fifth IEEE European Conference on Web Services (ECOWS 2007), 26–28 November 2007, Halle (Saale), pp. 171–180 (2007)
162. Etcheverry, L., Vaisman, A.A.: Enhancing OLAP analysis with Web cubes. In: Extended Semantic Web Conference, Halle (Saale), pp. 469–483 (2012)
163. Euzenat, J., Shvaiko, P.: Ontology Matching. Springer, Berlin (2007)

164. Ezenwoye, O., Sadjadi, S.M.: TRAP/BPEL: A framework for dynamic adaptation of composite services. In: Proceedings of the International Conference on Web Information Systems and Technologies (WEBIST 2007), Barcelona (2007)

165. Fahland, D., Favre, C., Koehler, J., Lohmann, N., Völzer, H., Wolf, K.: Analysis on demand: Instantaneous soundness checking of industrial business process models. Data Knowl. Eng. **70**(5), 448–466 (2011)

166. Fischer, L.: Workflow Handbook, 2002. Future Strategies, Lighthouse Point (2002)

167. Forgy, C.L.: Rete: A fast algorithm for the many pattern/many object pattern match problem. Artif. Intell. **19**(1), 17–37 (1982)

168. Franklin, M.J., Halevy, A.Y., Maier, D.: From databases to dataspaces: A new abstraction for information management. SIGMOD Rec. **34**(4), 27–33 (2005)

169. Fraternali, P., Brambilla, M., Vaca, C.: A model-driven approach to social bpm applications. In: Social BPM. Future Strategies, Lighthouse Point (2011)

170. Freire, J., Koop, D., Santos, E., Silva, C.T.: Provenance for computational tasks: A survey. Comput. Sci. Eng. **10**, 11–21 (2008)

171. Fritz, T., Murphy, G.C.: Using Information Fragments to Answer the Questions Developers Ask. ICSE'10, pp. 175–184. ACM, New York (2010)

172. Fuchs, M.: Adapting web services in a heterogeneous environment. In: International Conference on Web Services, San Diego, p. 656 (2004)

173. Furtado, C., Lima, A.A.B., Pacitti, E., Valduriez, P., Mattoso, M.: Physical and virtual partitioning in OLAP database clusters. In: SBAC-PAD, Rio de Janeiro, pp. 143–150 (2005)

174. Gartner: Business Activity Monitoring: Calm Before the Storm. ID Number: LE-15-9727 (2002)

175. Gater, A.: Process Matching and Discovery. Ph.D. thesis, University of Versailles (2012)

176. Gater, A., Grigori, D., Bouzeghoub, M.: Complex mapping discovery for semantic process model alignment. In: The 12th International Conference on Information Integration and Web-based Applications and Services, 8–10 November 2010, Paris, pp. 317–324. ACM, New York (2010)

177. Gater, A., Grigori, D., Bouzeghoub, M.: Owl-s process model matchmaking. In: IEEE International Conference on Web Services, ICWS 2010, Miami, 5–10 July 2010, pp. 640–641. IEEE Computer Society, Los Alamitos (2010)

178. Gater, A., Grigori, D., Haddad, M., Bouzeghoub, M., Kheddouci, H.: A summary-based approach for enhancing process model matchmaking. In: 2011 IEEE International Conference on Service-Oriented Computing and Applications, SOCA 2011, Irvine, 12–14 December 2011, SOCA'11, pp. 1–8 (2011)

179. Georgakopoulos, D., Hornick, M.F., Sheth, A.P.: An overview of workflow management: From process modeling to workflow automation infrastructure. Distrib. Parallel Databases **3**(2), 119–153 (1995)

180. Gerede, C.E., Su, J.: Specification and verification of artifact behaviors in business process models. In: International Conference on Service Oriented Computing, Vienna, pp. 181–192 (2007)

181. Glushko, R.J., McGrath, T.: Document Engineering. MIT, Cambridge (2005)

182. Goderis, A., Li, P., Goble, C.A.: Workflow discovery: The problem, a case study from e-science and a graph-based solution. In: 2006 IEEE International Conference on Web Services (ICWS 2006), 18–22 September 2006, Chicago, pp. 312–319 (2006)

183. Goderis, A., De Roure, D., Goble, C., Bhagat, J., Cruickshank, D., Fisher, P., Michaelides, D., Tanoh, F.: Discovering Scientific Workflows: The Myexperiment Benchmarks. Technical report, University of Southampton (2008)

184. Golfarelli, M., Rizzi, S., Proli, A.: Designing what-if analysis: Towards a methodology. In: Proceedings of ACM 9th International Workshop on Data Warehousing and OLAP (DOLAP 2006), Arlington, 10 November 2006, pp. 51–58. ACM, New York (2006) [ISBN 1-59593-530-4]

185. Gómez, L.I., Gómez, S.A., Vaisman, A.A.: A generic data model and query language for spatiotemporal OLAP cube analysis. In: Proceedings of 15th International Conference on

Extending Database Technology (EDBT '12), Berlin, 27–30 March 2012, pp. 300–311. ACM, New York (2012) [ISBN 978-1-4503-0790-1]

186. Google: Google desktop search. desktop.google.com (2004). Posted on 20 Feb 2015
187. Gottanka, R., Meyer, N.: ModelAsYouGo: (re-) design of S-BPM process models during execution time. In: S-BPM ONE Scientific Research. Lecture Notes in Business Information Processing, vol. 104, pp. 91–105. Springer, Berlin (2012)
188. Gouveia, L.: Multimedia Applications for Enterprise Information Systems. Ph.D. thesis, Master Dissertation, FEUP-DEEC, University of Oporto, Portugal (1994)
189. Grandi, F.: T-SPARQL: A TSQL2-like temporal query language for RDF. In: Local Proceedings of the Fourteenth East-European Conference on Advances in Databases and Information Systems, Novi Sad, 20–24 September 2010. CEUR Workshop Proceedings, vol. 639. CEUR-WS.org 2010
190. Graupner, S., Singhal, S., Basu, S., Motahari, H.: Enabling a Semantic Wiki to Drive Business Interactions. Technical Report: HPL-2009-193, HP Labs (2009)
191. Gray, J., Liu, D.T., Nieto-Santisteban, M.A., Szalay, A.S., DeWitt, D.J., Heber, G.: Scientific data management in the coming decade. SIGMOD Rec. **34**(4), 34–41 (2005)
192. Grebner, O., Ong, E., Riss, U.V., Brunzel, M., Bernardi, A., Roth-Berghofer, T.: Task Management Model - nepomuk Deliverable D3.1. Technical report (2006)
193. Greco, G., Guzzo, A., Pontieri, L.: Discovering expressive process models by clustering log traces. IEEE Trans. Knowl. Data Eng. **18**(8), 1010–1027 (2006)
194. Greenberg, S., Voida, S., Stehr, N., Tee, K.: Artifacts as instant messaging buddies. In: 43rd Hawaii International Conference on System Sciences (HICSS), 2010, pp. 1–10. IEEE, New York (2010)
195. Grigori, D., Casati, F., Dayal, U., Shan, M.-C.: Improving business process quality through exception understanding, prediction, and prevention. In: Proceedings of 27th International Conference on Very Large Data Bases, VLDB 2001, 11–14 September 2001, Roma, pp. 159–168 (2001)
196. Grigori, D., Casati, F., Castellanos, M., Dayal, U., Sayal, M., Shan, M.-C.: Business process intelligence. Comput. Ind. **53**(3), 321–343 (2004)
197. Grigori, D., Corrales, J.C., Bouzeghoub, M.: Behavioral matchmaking for service retrieval: Application to conversation protocols. The Journal of Information Systems (JIS) is the Academic Journal of the Accounting Information Systems (AIS) Section of the American Accounting Association (2008)
198. Grossman, R.L., Gu, Y.: Data mining using high performance data clouds: Experimental studies using sector and sphere. In: Proceedings of the 14th ACM SIGKDD International Conference on Knowledge Discovery and Data Mining, Las Vegas, 24–27 August 2008, pp. 920–927. ACM, New York (2008) [ISBN 978-1-60558-193-4]
199. Grund, M., Cudré-Mauroux, P., Madden, S.: A demonstration of HYRISE - a main memory hybrid storage engine. PVLDB **4**(12), 1434–1437 (2011)
200. Gualtieri, M., Rymer, J.R.: The forrester wave™: Complex event processing (cep) platforms, q3 2009. CEP (2009)
201. Guo, L., Shao, F., Botev, C., Shanmugasundaram, J.: XRANK: Ranked keyword search over XML documents. In: Special Interest Group on Management of Data Conference, San Diego, pp. 16–27 (2003)
202. Hacigümüs, H., Mehrotra, S., Iyer, B.R.: Providing database as a service. In: International Conference on Data Engineering, San Jose, pp. 29–38 (2002)
203. Hagel III, J., Brown, J.S.: From push to pull: Emerging models for mobilizing resources. J. Serv. Sci. (JSS) **1**(1), 93–110 (2011)
204. Halevy, A.Y.: Answering queries using views: A survey. VLDB J. **10**(4), 270–294 (2001)
205. Halevy, A.Y., Ashish, N., Bitton, D., Carey, M., Draper, D., Pollock, J., Rosenthal, A., Sikka, V.: Enterprise information integration: Successes, challenges and controversies. In: Proceedings of the ACM SIGMOD International Conference on Management of Data, Baltimore, 14–16 June 2005, pp. 778–787. ACM, New York (2005) [ISBN 1-59593-060-4]

206. Halevy, A.Y., Franklin, M.J., Maier, D.: Principles of dataspace systems. In: Proceedings of the Twenty-Fifth ACM SIGACT-SIGMOD-SIGART Symposium on Principles of Database Systems, 26–28 June 2006, Chicago, pp. 1–9. ACM, New York (2006) [ISBN 1-59593-318-2]

207. Hall, P.A.V., Dowling, G.R.: Approximate string matching. ACM Comput. Surv. **12**, 381–402 (1980)

208. Hall, J.M., Johnson, M.E.: When should a process be art, not science? Harv. Bus. Rev. **87**(3), 58–65 (2009)

209. Hallé, S., Villemaire, R.: Runtime monitoring of message-based workflows with data. In: 12th International IEEE Enterprise Distributed Object Computing Conference (ECOC 2008), 15–19 September 2008, Munich, pp. 63–72. IEEE Computer Society, Los Alamitos (2008) [ISBN 978-0-7695-3373-5]

210. Hallé, S., Villemaire, R.: XML methods for validation of temporal properties on message traces with data. In: On the Move to Meaningful Internet Systems: OTM 2008, OTM 2008 Confederated International Conferences, CoopIS, DOA, GADA, IS, and ODBASE 2008, Monterrey, 9–14 November 2008, Proceedings, Part I, pp. 337–353. Lecture Notes in Computer Science, vol. 5331. Springer, New York (2008) [ISBN 978-3-540-88870-3]

211. Hallows, J.E.: The project management office toolkit. AMACOM Div American Mgmt Assn. http://www.amazon.com/The-Project-Management-Office-Toolkit/dp/0814406637 (2002)

212. Hammoud, M., Rabbou, D.A., Nouri, R., Beheshti, S.-M.-R., Sakr, S.: DREAM: distributed RDF engine with adaptive query planner and minimal communication. PVLDB **8**(6), 654–665 (2015)

213. Han, J., Pei, J., Dong, G., Wang, K.: Efficient computation of iceberg cubes with complex measures. In: Special Interest Group on Management of Data Conference, Santa Barbara, pp. 1–12 (2001)

214. Han, J., Yan, X., Yu, P.S.: Scalable OLAP and mining of information networks. In: Proceedings of 12th International Conference on Extending Database Technology (EDBT 2009), Saint Petersburg, 24–26 March 2009. ACM International Conference Proceeding Series, vol. 360. ACM, New York (2009) [ISBN 978-1-60558-422-5]

215. Han, J., Sun, Y., Yan, X., Yu, P.S.: Mining knowledge from data: An information network analysis approach. In: International Conference on Data Engineering, Washington, DC (2012)

216. Harel, D., Naamad, A.: The statemate semantics of statecharts. ACM Trans. Softw. Eng. Methodol. (TOSEM) **5**(4), 293–333 (1996)

217. Harel, D., Politi, M.: Modeling Reactive Systems with Statecharts: The STATEMATE Approach. Computer Science and Information Processing. McGraw-Hill, New York (1998)

218. Hassanzadeh, O., Kementsietsidis, A., Lim, L., Miller, R.J., Wang, M.: A framework for semantic link discovery over relational data. In: Conference on Information and Knowledge Management, Hong Kong, pp. 1027–1036 (2009)

219. Hassanzadeh, O., Duan, S., Fokoue, A., Kementsietsidis, A., Srinivas, K., Ward, M.J.: Helix: Online enterprise data analytics. In: World Wide Web (Companion Volume), Hyderabad, pp. 225–228 (2011)

220. Hay, D., Healy, K.A.: Defining business rules-what are they really. Final Report (2000)

221. He, B., Chang, K.C.-C.: Statistical schema matching across web query interfaces. In: Proceedings of the 2003 ACM SIGMOD International Conference on Management of Data, SIGMOD '03, San Diego, pp. 217–228 (2003)

222. Henzinger, T.A., Qadeer, S., Rajamani, S.K., Tasiran, S.: An assume-guarantee rule for checking simulation. ACM Trans. Program. Lang. Syst. **24**(1), 51–64 (2002)

223. Hepp, M., Leymann, F., Domingue, J., Wahler, A., Fensel, D.: Semantic business process management: A vision towards using semantic web services for business process management. In: International Conference on e-Business Engineering, Beijing, pp. 535–540 (2005)

224. Herbst, J.: A machine learning approach to workflow management. In: European Conference on Machine Learning, Barcelona, pp. 183–194 (2000)

225. Hernández, M.A., Stolfo, S.J.: Real-world data is dirty: Data cleansing and the merge/purge problem. Data Min. Knowl. Discov. **2**(1), 9–37 (1998)

226. Holland, D.A., Braun, U., Maclean, D., Muniswamy-Reddy, K.K., Seltzer, M.: Choosing a data model and query language for provenance. In: Second International Provenance and Annotation Workshop, University of Utah, Salt Lake City, 17–18 June 2008

227. Hollingsworthm, D.: The Workflow Reference Model. Workflow Management Coalition. http://www.wfmc.org/standards/docs/tc003v11.pdf (1995)

228. Holme, P., Sarami, J.: Temporal networks. CoRR, abs/1108.1780 (2011)

229. Holz, H., Rostanin, O., Dengel, A., Suzuki, T., Maeda, K., Kanasaki, K.: Task-based process know-how reuse and proactive information delivery in TaskNavigator. In: Conference on Information and Knowledge Management, Arlington, pp. 522–531 (2006)

230. Horrocks, I., Patel-Schneider, P.F., van Harmelen, F.: From shiq and rdf to owl: The making of a web ontology language. J. Web Semant. 1(1), 7–26 (2003)

231. Hull, R.: Artifact-centric business process models: Brief survey of research results and challenges. In: OTM Conferences (2), Monterrey, pp. 1152–1163 (2008)

232. Hull, R., Narendra, N.C., Nigam, A.: Facilitating workflow interoperation using artifact-centric hubs. In: Service-Oriented Computing, pp. 1–18. Springer, Berlin (2009)

233. Hull, R., Damaggio, E., Fournier, F., Gupta, M., Heath III, F.T., Hobson, S., Linehan, M., Maradugu, S., Nigam, A., Sukaviriya, P., et al.: Introducing the guard-stage-milestone approach for specifying business entity lifecycles. In: Web Services and Formal Methods, pp. 1–24. Springer, Berlin (2011)

234. Husain, M.F., Doshi, P., Khan, L., Thuraisingham, B.M.: Storage and retrieval of large RDF graph using Hadoop and MapReduce. In: Proceedings of the Cloud Computing, First International Conference (CloudCom 2009), Beijing, 1–4 December 2009, pp. 680–686. Lecture Notes in Computer Science, vol. 5931. Springer, Berlin (2009) [ISBN 978-3-642-10664-4]

235. Husain, M.F., Khan, L., Kantarcioglu, M., Thuraisingham, B.M.: Data intensive query processing for large RDF graphs using cloud computing tools. In: IEEE International Conference on Cloud Computing (CLOUD 2010), Miami, 5–10 July 2010, pp. 1–10. IEEE, New York (2010) [ISBN 978-1-4244-8207-8]

236. Indulska, M., Recker, J., Rosemann, M., Green, P.F.: Business process modeling: Current issues and future challenges. In: Conference on Advanced Information Systems Engineering, Amsterdam, pp. 501–514 (2009)

237. Ives, Z.G., Khandelwal, N., Kapur, A., Cakir, M.: ORCHESTRA: Rapid, collaborative sharing of dynamic data. In: Conference on Innovative Data Systems Research, Asilomar, pp. 107–118 (2005)

238. JBossDrools: Drools expert user guide, version 5.5 (2015). https://docs.jboss.org/drools/release/5.5.0.Final/drools-expert-docs/html_single/. Posted on 20 Feb 2015

239. JBossDrools: Drools fusion user guide, version 5.5 (2015). https://docs.jboss.org/drools/release/5.5.0.Final/drools-fusion-docs/pdf/drools-fusion-docs.pdf. Posted on 20 Feb 2015

240. Jennings, B., Finkelstein, A.: Micro workflow gestural analysis: Representation in social business processes. In: Business Process Management Workshops, pp. 278–290. Springer, Berlin (2010)

241. Jin, T., Wang, J., Wen, L.: Efficient retrieval of similar business process models based on structure. In: OTM Conferences, Hersonissos, pp. 56–63 (2011)

242. Jin, T., Wang, J., Wen, L.: Efficiently querying business process models with BeehiveZ. In: Proceedings of the Demo Track of the Nineth Conference on Business Process Management (BPM Demos), Clermont-Ferrand (2011)

243. Jin, T., Wang, J., Wen, L.: Querying business process models based on semantics. In: 16th International Conference on Database Systems for Advanced Applications (DASFAA 2011), Hong Kong, 22–25 April 2011, Proceedings, Part II. Lecture Notes in Computer Science, vol. 6588, pp. 164–178. Springer, Berlin (2011) [ISBN 978-3-642-20151-6]

244. Jin, T., Wang, J., La Rosa, M., ter Hofstede, A.H.M., Wen, L.: Efficient querying of large process model repositories. Comput. Ind. **64**(1), 41–49 (2013)

245. Johannesson, P., Andersson, B., Wohed, P.: Business process management with social software systems—a new paradigm for work organisation. In: Business Process Management Workshops, pp. 659–665. Springer, Berlin (2009)
246. Johnson, B.C., Manyika, J.M., Yee, L.A.: The next revolution in interactions. McKinsey Q. **4**, 20–33 (2005)
247. Jorgensen, H.D.: Interactive process models. Ph.D. Thesis, Norwegian University of Science and Technology, Trondheim, Norway (2004)
248. Kämpgen, B., Harth, A.: Transforming statistical linked data for use in OLAP systems. In: Proceedings of the 7th International Conference on Semantic Systems - I-SEMANTICS, Graz, pp. 33–40 (2011)
249. Kandel, S., Paepcke, A., Theobald, M., Garcia-Molina, H., Abelson, E.: Photospread: A spreadsheet for managing photos. In: Proceedings of the SIGCHI Conference on Human Factors in Computing Systems, pp. 1749–1758. ACM, New York (2008)
250. Kang, U., Chau, D.H., Faloutsos, C.: Pegasus: Mining billion-scale graphs in the cloud. In: IEEE International Conference on Acoustics, Speech and Signal Processing (ICASSP), pp. 5341–5344. IEEE, New York (2012)
251. Kaptelinin, V., Czerwinski, M.: Beyond the Desktop Metaphor: Designing Integrated Digital Work Environments, vol. 1. MIT, Cambridge (2007)
252. Karger, D.R., Bakshi, K., Huynh, D., Quan, D., Sinha, V.: Haystack: A general-purpose information management tool for end users based on semistructured data. In: Conference on Innovative Data Systems Research, Asilomar, pp. 13–26 (2005)
253. Karvounarakis, G., Ives, Z.G., Tannen, V.: Querying data provenance. In: Proceedings of the Special Interest Group on Management of Data. ACM, New York (2010)
254. Kaschner, K., Wolf, K.: Set algebra for service behavior: Applications and constructions. In: Proceedings of the 7th International Conference on Business Process Management (BPM), Ulm, 8–10 September 2009, pp. 193–210 (2009)
255. Kawamura, T., De Blasio, J.A., Hasegawa, T., Paolucci, M., Sycara, K.: A preliminary report of a public experiment of a semantic service matchmaker combined with a uddi business registry. In: International Conference on Service Oriented Computing (ICSOC), Trento (2003)
256. Kelly W.: Social Task Management vs. Project Management Tools for Your SMB. White paper. Techrepublic (2013). http://www.techrepublic.com/blog/it-consultant/social-task-management-vs-project-management-tools-for-your-smb/
257. Kemper, A., Neumann, T.: HyPer: A hybrid OLTP&OLAP main memory database system based on virtual memory snapshots. In: International Conference on Data Engineering (ICDE), Hannover, pp. 195–206 (2011)
258. Kephart, J.O., Chess, D.M.: The vision of autonomic computing. IEEE Comput. **36**(1), 41–50 (2003)
259. Kiefer, C., Bernstein, A., Lee, H.J., Klein, M., Stocker, M.: Semantic process retrieval with isparql. In: Extended Semantic Web Conference (ESWC), pp. 609–623 Springer, Berlin (2007)
260. Kiefer, C., Bernstein, A., Stocker, M.: The fundamentals of isparql: A virtual triple approach for similarity-based semantic web tasks. In: The Semantic Web. 6th International Semantic Web Conference, 2nd Asian Semantic Web Conference (ISWC 2007 + ASWC 2007), Busan, 11–15 November 2007. Lecture Notes in Computer Science, vol. 4825, pp. 295–309. Springer, Berlin (2007) [ISBN 978-3-540-76297-3]
261. Kim, H., Ravindra, P., Anyanwu, K.: From SPARQL to MapReduce: The journey using a nested triplegroup algebra. Proc. VLDB Endowment **4**(12), 1426–1429 (2011)
262. Kipgen, B., O'Riain, S., Harth, A.: Interacting with statistical linked data via OLAP operations. In: Proceedings of Interacting with Linked Data-Extended Semantic Web Conference, Heraklion (2012)
263. Klinkmüller, C., Weber, I., Mendling, J., Leopold, H., Ludwig, A.: Increasing recall of process model matching by improved activity label matching. In: Proceedings of the 11th International Conference on Business Process Management (BPM), 26–30 August 2013, Beijing, pp. 211–218 (2013)

264. Kofman, A., Yaeli, A., Klinger, T., Tarr, P.: Roles, rights, and responsibilities: Better governance through decision rights automation. In: Proceedings of the 2009 ICSE Workshop on Software Development Governance, pp. 9–14. IEEE Computer Society, Los Alamitos (2009)

265. Kohavi, R., Rothleder, N.J., Simoudis, E.: Emerging trends in business analytics. Commun. ACM 45(8), 45–48 (2002)

266. Kongdenfha, W., Benatallah, B., Saint-Paul, R., Casati, F.: Spreadmash: A spreadsheet-based interactive browsing and analysis tool for data services. In: Advanced Information Systems Engineering, pp. 343–358. Springer, Berlin (2008)

267. Köpcke, H., Thor, A., Rahm, E.: Evaluation of entity resolution approaches on real-world match problems. Proc. VLDB Endowment 3(1), 484–493 (2010)

268. Kostakos, V.: Temporal graph. Phys. A Stat. Mech. Appl. 388(6), 1007–1023 (2009)

269. Koutsoukis, N.S., Mitra, G., Lucas, C.: Adapting on-line analytical processing for decision modelling: The interaction of information and decision technologies. Decis. Support Syst. 26(1), 1–30 (1999)

270. Kraut, R., Egido, C., Galegher, J.: Patterns of contact and communication in scientific research collaboration. In: Proceedings of the 1988 ACM Conference on Computer-Supported Cooperative Work, pp. 1–12. ACM, New York (1988)

271. Kritikos, K., Plexousakis, D.: Semantic qos metric matching. In: European Conference on Web Services, Zürich, pp. 265–274 (2006)

272. Kritikos, K., Plexousakis, D.: Requirements for qos-based web service description and discovery. IEEE Trans. Serv. Comput. 2(4), 320–337 (2009)

273. Kritikos, K., Pernici, B., Plebani, P., Cappiello, C., Comuzzi, M., Benbernou, S., Brandic, I., Kertész, A., Parkin, M., Carro, M.: A survey on service quality description. ACM Comput. Surv. 46(1), 1 (2013)

274. Kumaran, S., Liu, R., Wu, F.Y.: On the duality of information-centric and activity-centric models of business processes. In: Advanced Information Systems Engineering, pp. 32–47. Springer, Berlin (2008)

275. Kunze, M., Weske, M.: Metric trees for efficient similarity search in large process model repositories. In: Business Process Management Workshops - BPM 2010 International Workshops and Education Track, Hoboken, 13–15 September 2010, Revised Selected Papers. Lecture Notes in Business Information Processing, vol. 66, pp. 535–546. Springer, New York (2011) [ISBN 978-3-642-20510-1]

276. Kunze, M., Weidlich, M., Weske, M.: Behavioral similarity - a proper metric. In: Proceedings of the 9th International Conference on Business Process Management (BPM), 30 August–2 September 2011, pp. 166–181, Clermont-Ferrand (2011)

277. Kurniawan, T.A., Ghose, A.K., Li, L.S., Dam, H.K.: On formalizing inter-process relationships. In: Business Process Management Workshops. Lecture Notes in Business Information Processing, vol. 100, pp. 75–86. Springer, Berlin (2012)

278. Küster, J.M., Ryndina, K., Gall, H.: Generation of business process models for object life cycle compliance. In: Business Process Management, pp. 165–181. Springer, Berlin (2007)

279. Küster, J.M., Gerth, C., Förster, A., Engels, G.: Detecting and resolving process model differences in the absence of a change log. In: 6th International Conference on Business Process Management (BPM), 2–4 September 2008. Lecture Notes in Computer Science, pp. 244–260. Springer, New York (2008)

280. Langford, J., Li, L., Strehl, A.: Vowpal wabbit. https://github.com/JohnLangford/vowpal_wabbit/wiki (2011)

281. La Rosa, M., Dumas, M., Uba, R., Dijkman, R.M.: Business process model merging: An approach to business process consolidation. ACM Trans. Softw. Eng. Methodol. 22(2), 11 (2013)

282. Lausen, H., Polleres, A., Roman, D.: Web Service Modeling Ontology (WSMO). W3C Submission. http://www.w3.org/Submission/WSMO/ (2005)

283. Le Clair, C., Moore, C.: Dynamic Case Management: An Old Idea Catches New Fire. Forrester Research, Cambridge (2009)

284. Leavitt, N.: Are web services finally ready to deliver? IEEE Comput. **37**(11), 14–18 (2004)
285. Lee, Y., Sayyadian, M., Doan, A., Rosenthal, A.S.: Etuner: Tuning schema matching software using synthetic scenarios. VLDB J. **16**(1), 97–122 (2007)
286. Lemos, F., Grigori, D., Bouzeghoub, M.: In: Proceedings of the 12th International Conference on Web Engineering (ICWE), Berlin, 23–27 July 2012, pp. 299–306 (2012)
287. Levenshtein, V.I.: Binary codes capable of correcting deletions, insertions and reversals. Sov. Phys. Dokl. **10**, 707–710 (1966)
288. Levy, A.Y., Rajaraman, A., Ordille, J.J.: Querying heterogeneous information sources using source descriptions. In: Proceedings of 22th International Conference on Very Large Data Bases (VLDB), 3–6 September 1996, pp. 251–262. Morgan Kaufmann, Los Altos (1996)
289. Leymann, F., Roller, D.: Production Workflow: Concepts and Techniques. Prentice Hall PTR, Upper Saddle River (2000)
290. Leymann, F., Roller, D., Schmidt, M.-T.: Web services and business process management. IBM Syst. J. **41**(2), 198–211 (2002)
291. Li, L., Horrocks, I.: A software framework for matchmaking based on semantic web technology. In: International World Wide Web Conference (WWW), Budapest, pp. 331–339 (2003)
292. Li, W.-S., Clifton, C.: Semint: A tool for identifying attribute correspondences in heterogeneous databases using neural networks. Data Knowl. Eng. **33**(1), 49–84 (2000)
293. Li, W.-S., Clifton, C., Liu, S.-Y.: Database integration using neural networks: Implementation and experiences. Knowl. Inf. Syst. **2**(1), 73–96 (2000)
294. Li, C., Reichert, M., Wombacher, A.: On measuring process model similarity based on high-level change operations. In: 27th International Conference on Conceptual Modeling- ER 2008, 20–24 October 2008. Lecture Notes in Computer Science, pp. 248–264. Springer, New York (2008)
295. Li, J., Tang, J., Li, Y., Luo, Q.: Rimom: A dynamic multistrategy ontology alignment framework. IEEE Trans. Knowl. Data Eng. **21**(8), 1218–1232 (2009)
296. Lim, L., Wang, H., Wang, M.: Semantic queries in databases: Problems and challenges. In: Conference on Information and Knowledge Management (CIKM), Hong Kong, pp. 1505–1508 (2009)
297. Lima, A.A.B., Mattoso, M., Valduriez, P.: Adaptive virtual partitioning for OLAP query processing in a database cluster. J. Inf. Data Manag. **1**(1), 75–88 (2010)
298. Lin, D.: An information-theoretic definition of similarity. In: International Machine Learning Society (ICML), Wisconsin, pp. 296–304 (1998)
299. Liu, D.R., Shen, M.: Workflow modeling for virtual processes: An order-preserving process-view approach. Inf. Syst. **28**(6), 505–532 (2003)
300. Liu, R., Bhattacharya, K., Wu, F.Y.: Modeling business contexture and behavior using business artifacts. In: Advanced Information Systems Engineering, pp. 324–339. Springer, New York (2007)
301. Low, Y., Gonzalez, J.E., Kyrola, A., Bickson, D., Guestrin, C.E., Hellerstein, J.: Graphlab: A new framework for parallel machine learning (2014). Preprint. arXiv:1408.2041
302. Lu, R., Sadiq, S.: On the discovery of preffered work practice throught business process variants. In: ER 2007. Springer, New York (2007)
303. Luckham, D.: The power of events: An introduction to complex event processing in distributed enterprise systems. In: Rule Representation Interchange and Reasoning on the Web. Springer, New York (2008)
304. Ly, L.T., Knuplesch, D., Rinderle-Ma, S., Göser, K., Pfeifer, H., Reichert, M., Dadam, P.: SeaFlows toolset - compliance verification made easy for process-aware information systems. In: Conference on Advanced Information Systems Engineering Forum 2010, Hammamet, pp. 76–91 (2010)
305. Ly, L.T., Maggi, F.M., Montali, M., Rinderle-Ma, S., van der Aalst, W.M.P.: Compliance monitoring in business processes: Functionalities, application, and tool-support. Inf. Syst. **54**, 209–234 (2015)
306. Lynch, C.: Big data: How do your data grow? Nature **455**(7209), 28–29 (2008)

307. Maamar, Z., Sakr, S., Barnawi, A., Beheshti, S.-M.-R.: A framework of enriching business processes life-cycle with tagging information. In: Databases Theory and Applications - Proceedings of the 26th Australasian Database Conference (ADC), Melbourne, 4–7 June 2015, pp. 309–313 (2015)

308. Mabrouk, N.B., Beauche, S., Kuznetsova, E., Georgantas, N., Issarny, V.: Qos-aware service composition in dynamic service oriented environments. In: Middleware, pp. 123–142. Springer, New York (2009)

309. Madhavan, J., Bernstein, P.A., Rahm, E.: Generic schema matching with cupid. In: Proceedings of the 27th International Conference on Very Large Data Bases, Roma, pp. 49–58 (2001)

310. Madhavan, J., Bernstein, P.A., Doan, A., Halevy, A.: Corpus-based schema matching. In: Proceedings of the 21st International Conference on Data Engineering, Tokyo, pp. 57–68 (2005)

311. Maedche, A., Staab, S.: Measuring similarity between ontologies. In: Proceedings of the 13th International Conference on Knowledge Engineering and Knowledge Management. Ontologies and the Semantic Web, Siguenza, pp. 251–263 (2002)

312. Maggi, F.M., Montali, M., Westergaard, M., van der Aalst, W.M.P.: Monitoring business constraints with linear temporal logic: An approach based on colored automata. In: Proceedings of the 9th International Conference on Business Process Management (BPM), Clermont-Ferrand, 30 August–2 September, 2011, pp. 132–147 (2011)

313. Mahbub, K., Spanoudakis, G.: A framework for requirents monitoring of service based systems. In: International Conference on Service Oriented Computing (ICSOC), New York, pp. 84–93 (2004)

314. Malewicz, G., Austern, M.H., Bik, A.J.C., Dehnert, J.C., Horn, I., Leiser, N., Czajkowski, G.: Pregel: A system for large-scale graph processing. In: Special Interest Group on Management of Data Conference, Indianapolis, pp. 135–146 (2010)

315. Manola, F., Miller, E.: RDF Primer. W3C. http://www.w3.org/TR/rdf-primer/ (2004)

316. Marmel, E.: Microsoft Project 2007 Bible, vol. 767. Wiley, New York (2011)

317. Martín-Díaz, O., Cortés, A.R., Benavides, D., Durán, A., Toro, M.: A quality-aware approach to web services procurement. In: Technologies for E-Services (TES), pp. 42–53. Springer, New York (2003)

318. Mathiesen, P., Watson, J., Bandara, W., Rosemann, M.: Applying social technology to business process lifecycle management. In: Business Process Management Workshops. Lecture Notes in Business Information Processing, vol. 99, pp. 231–241. Springer, New York (2012)

319. Matz, O., Miller, G.A., Potthoff, A., Thomas, W., Valkema, E.: Report on the program amore. Technical Report, Institut fur Informatik und Praktische Mathematik der Christian-Albrechts-Universitat zu Kiel (1995)

320. Maximilien, E.M., Singh, M.P.: Conceptual model of web service reputation. SIGMOD Rec. **31**(4), 36–41 (2002)

321. May, W., Schenk, F., von Lienen, E.: Extending an owl web node with reactive behavior. In: Principles and Practice of Semantic Web Reasoning, pp. 134–148. Springer, New York (2006)

322. Mecca, G., Papotti, P., Raunich, S.: Core schema mappings: Scalable core computations in data exchange. Inf. Syst. **37**(7), 677–711 (2012)

323. Medeiros, A.K.A., van der Aalst, W.M.P., Weijters, A.J.M.M.: Quantifying process equivalence based on observed behavior. Data Knowl. Eng. **64**(1), 55–74 (2008)

324. Medeiros, A.K.A.D., van der Aalst, W.M.P., Pedrinaci, C.: Semantic process mining tools: Core building blocks. In: The European Conference on Information Systems (ECIS), Galway, pp. 1953–1964 (2008)

325. Melnik, S., Garcia-Molina, H., Rahm, E.: Similarity flooding: A versatile graph matching algorithm and its application to schema matching. In: Proceedings of the 18th International Conference on Data Engineering, San Jose (2002)

326. Mendling, J.: Metrics for Process Models. Event-Driven Process Chains (EPC), pp. 17–57. Lecture Notes in Business Information Processing. Springer, New York (2009)

327. Mendling, J., Lassen, K.B., Zdun, U.: Transformation strategies between block-oriented and graph-oriented process modelling languages. In: Lehner, F., Niekabel, H., Kleinschmidt, P. (eds.) Multikonferenz Wirtschaftsinformatik, pp. 297–312. GITO, Berlin (2006)
328. Michael, M.: Cloud Computing. Pearson Education, New Delhi (2009)
329. Microsoft Corporation: XLANG: Web services for business process design (2001). https://msdn.microsoft.com/en-us/library/aa577463.aspx. Posted on 20 Feb 2015
330. Milanovic, M., Gasevic, D., Wagner, G.: Combining rules and activities for modeling service-based business processes. In: Twelfth International Enterprise Distributed Object Computing Conference Workshops, pp. 11–22. IEEE, New York (2008)
331. Milen, M., Shapiro, R.: Business process analytics. In: Handbook on Business Process Management 2. International Handbooks on Information Systems, pp. 137–157. Springer, New York (2010)
332. Mili, H., Tremblay, G., Jaoude, G.B., Lefebvre, É., Elabed, L., Boussaidi, G.E.: Business process modeling languages: Sorting through the alphabet soup. ACM Comput. Surv. (CSUR) **43**(1), 4 (2010)
333. Miller, G.: Wordnet: A lexical database for english. Commun. ACM **38**(11), 39–41 (1995)
334. Milner, R.: Communication and Concurrency. Prentice-Hall, London (1989)
335. Milo, T., Zohar, S.: Using schema matching to simplify heterogeneous data translation. In: Proceedings of the 24rd International Conference on Very Large Data Bases, New York, pp. 122–133 (1998)
336. Minor, M., Tartakovski, A., Bergmann, R.: Representation and structure-based similarity assessment for agile workflows. In: Proceedings of the 7th International Conference on Case-Based Reasoning, Halle (2007)
337. Mitra, N., Lafon, Y.: Simple Object Access Protocol (SOAP) 1.2. W3C Recommendation. http://www.w3.org/TR/soap/ (2007)
338. Mitsa, T.: Temporal Data Mining, 1st edn. Chapman & Hall/CRC, London (2010)
339. Momotko, M., Subieta, K.: Process query language: A way to make workflow processes more flexible. In: Proceesing of 8th East European Conference on Advances in Databases and Information Systems (ADBIS 2004), Budapest, 22–25 September 2004. Lecture Notes in Computer Science, vol. 3255. Springer, New York (2004) [ISBN 3-540-23243-5]
340. Montali, M., Maggi, F.M., Chesani, F., Mello, P., van der Aalst, W.M.P.: Monitoring business constraints with the event calculus. ACM Trans. Intell. Syst. Technol. **5**(1), 17 (2013)
341. Moreau, L.: Provenance-based reproducibility in the semantic Web. J. Web Semant. **9**(2), 202–221 (2011)
342. Moreau, L., Freire, J., Futrelle, J., Mcgrath, R.E., Myers, J., Paulson, P.: The open provenance model: An overview. In: International Provenance and Annotation Workshop (IPAW), pp. 323–326. Springer, New York (2008)
343. Motahari-Nezhad, H.R., Bartolini, C.: Next best step and expert recommendation for collaborative processes in it service management. In: Business Process Management (BPM), pp. 50–61. Springer, New York (2011)
344. Motahari-Nezhad, H.R., Benatallah, B., Saint-Paul, R., Casati, F., Andritsos, P.: Process spaceship: Discovering and exploring process views from event logs in data spaces. Proc. VLDB Endowment **1**(2), 1412–1415 (2008)
345. Motahari-Nezhad, H.R., Saint-Paul, R., Benatallah, B., Casati, F.: Deriving protocol models from imperfect service conversation logs. IEEE Trans. Knowl. Data Eng. **20**, 1683–1698 (2008)
346. Motahari-Nezhad, H.R., Saint-Paul, R., Casati, F., Benatallah, B.: Event correlation for process discovery from web service interaction logs. VLDB J. **20**(3), 417–444 (2011)
347. Muehlen, M., Rosemann, M.: Workflow-based process monitoring and controlling - technical and organizational issues. In: The Hawaii International Conference on System Sciences (HICSS), Maui (2000)
348. Muehlen, M.Z.: Workflow-based process controlling. Foundation, Design, and Application of Workflow-Driven Process Information Systems. Logos, Berlin (2004)

349. Mulo, E., Zdun, U., Dustdar, S.: Monitoring web service event trails for business compliance. In: IEEE International Conference on Service-Oriented Computing and Applications (SOCA), Taipei, pp. 1–8 (2009)
350. Mulo, E., Zdun, U., Dustdar, S.: Domain-specific language for event-based compliance monitoring in process-driven SOAs. Serv. Oriented Comput. Appl. **7**(1), 59–73 (2013)
351. Muñoz-Gama, J., Carmona, J.: Enhancing precision in process conformance: Stability, confidence and severity. In: Proceedings of the IEEE Symposium on Computational Intelligence and Data Mining (CIDM 2011). Part of the IEEE Symposium Series on Computational Intelligence 2011, 11–15 April 2011, Paris, pp. 184–191. IEEE, New York (2011) [ISBN 978-1-4244-9925-0]
352. Murata, T.: Petri nets: Properties, analysis and applications. Proc. IEEE **77**(4), 541–580 (1989). NewsletterInfo: 33Published as Proc. IEEE, New York **77**(4)
353. Nejati, S., Sabetzadeh, M., Chechik, M., Easterbrook, S., Zave, P.: Matching and merging of statecharts specifications. In: Proceedings of the 29th International Conference on Software Engineering, Minneapolis (2007)
354. Neumeyer, L., Robbins, B., Nair, A., Kesari, A.: S4: distributed stream computing platform. In: The 10th IEEE International Conference on Data Mining Workshops (ICDMW), Sydney, 13 December 2010, pp. 170–177 (2010)
355. Newman, M.: Small Worlds: The Dynamics of Networks Between Order and Randomness. Oxford University Press, Oxford (2010)
356. Nezhad, H.R.M., Benatallah, B., Martens, A., Curbera, F., Casati, F.: Semi-automated adaptation of service interactions. In: Proceedings of the 16th International Conference on World Wide Web (WWW), Banff, 8–12 May 2007, pp. 993–1002 (2007)
357. Nezhad, H.R.M., Xu, G.Y., Benatallah, B.: Protocol-aware matching of web service interfaces for adapter development. In: Proceedings of the 19th International Conference on World Wide Web (WWW), Raleigh, 26–30 April 2010, pp. 731–740 (2010)
358. Nigam, A., Caswell, N.S.: Business artifacts: An approach to operational specification. IBM Syst. J. **42**(3), 428–445 (2003)
359. ObjectManagementGroup(OMG): Semantics of business vocabulary and business rules (SBVR) (2015). http://www.omg.org/spec/SBVR/. Posted on 20 Feb 2015
360. Oldham, N., Verma, K., Sheth, A., Hakimpour, F.: Semantic ws-agreement partner selection. In: Proceedings of the 15th International Conference on World Wide Web (WWW), Edinburgh, pp. 23–26 (2006)
361. Olding, E.: Social BPM: Getting to doing. Technical Report (2011)
362. Olding, E., Rozwell, C.: Expand your BPM horizons by exploring unstructured processes. Technical Report (2009)
363. Olding, E., Rozwell, C., Sinur, J.: Social BPM: Design by doing. Technical Report (2010)
364. Olston, C., Reed, B., Srivastava, U., Kumar, R., Tomkins, A.: Pig latin: A not-so-foreign language for data processing. In: Proceedings of the 2008 ACM SIGMOD International Conference on Management of Data, pp. 1099–1110. ACM, New York (2008)
365. Object Management Group (OMG): OMG Final Adopted Specification: Business Process Modeling Notation Specification (2006). http://www.omg.org/spec/. Posted on 20 Feb 2015
366. Object Management Group (OMG): Case Management Model and Notation (CMMN) (2014). http://www.omg.org/spec/CMMN/. Posted on 20 Feb 2015
367. Oracle: Exploiting the power of Oracle using Microsoft Excel. http://www.oracle.com/technology/products/bi/pdf/BI_Spreadsheet_Addin_WP.pdf (2004)
368. O'Reilly, T.: What is web 2.0 - design patterns and business models for the next generation of software (2005) [cited January 2012]
369. Ouyang, C., Wynn, M.T., Fidge, C., ter Hofstede, A.H.M., Kuhr, J.C.: Modelling complex resource requirements in business process management systems. In: 21st Australasian Conference on Information Systems: Defining and Establishing a High Impact Discipline (ACIS 2010), Queensland University of Technology, Brisbane, December 2010
370. Papazoglou, M.P., Heuvel, W.J.V.D.: Service oriented architectures: Approaches, technologies and research issues. VLDB J. **16**(3), 389–415 (2007)

371. Paredaens, J., Peelman, P., Tanca, L.: G-log: A graph-based query language. IEEE Trans. Knowl. Data Eng. **7**(3), 436–453 (1995)
372. Paschke, A.: The reaction ruleml classification of the event/action/state processing and reasoning space (2006). arXiv preprint cs/0611047
373. Paschke, A., Kozlenkov, A.: Rule-based event processing and reaction rules. In: Rule Interchange and Applications, pp. 53–66. Springer, New York (2009)
374. Patil, A., Oundhakar, S., Sheth, A., Verna, K.: Meteor-s web service annotation framework. In: World Wide Web Conference, Manhattan (2004)
375. Pedrinaci, C., Domingue, J., Medeiros, A.K.A.D.: A core ontology for business process analysis. In: European Semantic Web Conference (ESWC), pp. 49–64. Springer, New York (2008)
376. Pegarules: https://www-304.ibm.com/partnerworld/gsd/solutiondetails.do?solution=14507& lc=en. Accessed July 2015
377. Pesic, M., Schonenberg, H., van der Aalst, W.M.P.: DECLARE: full support for loosely-structured processes. In: 11th IEEE International Enterprise Distributed Object Computing Conference (EDOC 2007), 15–19 October 2007, Annapolis. IEEE Computer Society, Los Alamitos (2007) [ISBN 0-7695-2891-0]
378. Pesic, M., Schonenberg, M.H., Sidorova, N., van der Aalst, W.M.P.: Constraint-based workflow models: Change made easy. In: On the Move to Meaningful Internet Systems 2007: CoopIS, DOA, ODBASE, GADA, and IS, pp. 77–94. Springer, New York (2007)
379. Peterson, J.L.: Petri Net Theory and the Modeling of Systems. Prentice Hall PTR, Upper Saddle River (1981)
380. Petit, J.M., Toumani, F., Boulicaut, J.F., Kouloumdjian, J.: Towards the reverse engineering of denormalized relational databases. In: International Conference on Data Engineering (ICDE), New Orleans, pp. 218–227 (1996)
381. Pike, R., Dorward, S., Griesemer, R., Quinlan, S.: Interpreting the data: Parallel analysis with sawzall. Sci. Program. **13**(4), 277–298 (2005)
382. Plattner, H.: A common database approach for OLTP and OLAP using an in-memory column database. In: Proceedings of the 2009 ACM SIGMOD International Conference on Management of Data (SIGMOD), pp. 1–2. ACM, New York (2009)
383. Pomello, L., Rozenberg, G., Simone, C.: A survey of equivalence notions for net based systems. In: Advances in Petri Nets: The DEMON Project. Lecture Notes in Computer Science, vol. 609, pp. 410–472. Springer, Heidelberg (1992)
384. Ponge, J., Benatallah, B., Casati, F., Toumani, F.: Analysis and applications of timed service protocols. ACM Trans. Softw. Eng. Methodol. **19**(4), 1–38 (2010)
385. Ponnekanti, S., Fox, A.: Interoperability among independently evolving web services. In: Middleware, pp. 331–351. Springer, Heidelberg (2004)
386. Porter, M.F.: Readings in information retrieval. An Algorithm for Suffix Stripping, pp. 313–316. Morgan Kaufmann, Los Altos (1997)
387. Qian, T., Yang, Y., Wang, S.: Refining graph partitioning for social network clustering. In: Web Information Systems Engineering (WISE), pp. 77–90. Springer, Heidelberg (2010)
388. Qu, Q., Zhu, F., Yan, X., Han, J., Yu, P.S., Li, H.: Efficient topological OLAP on information networks. In: Proceedings of the Database Systems for Advanced Applications - 16th International Conference (DASFAA 2011), Part I, Hong Kong, 22–25 April 2011. Lecture Notes in Computer Science, vol. 6587. Springer, Berlin (2011) [ISBN 978-3-642-20148-6]
389. R Core Team: R: A Language and Environment for Statistical Computing. R Foundation for Statistical Computing (2012/2014)
390. Rahm, E., Bernstein, P.A.: A survey of approaches to automatic schema matching. VLDB J. **10**(4), 334–350 (2001)
391. Rahm, E., Do, H.H.: Data cleaning: Problems and current approaches. IEEE Data Eng. Bull. **23**(4), 3–13 (2000)
392. Ran, S.: A model for web services discovery with qos. ACM SIGecom Exch. **4**, 1–10 (2003)
393. Redding, G., Dumas, M., ter Hofstede, A.H.M., Iordachescu, A.: Generating business process models from object behavior models. Inf. Syst. Manag. **25**(4), 319–331 (2008)

394. Reichert, M., Dadam, P.: Adept$_{flex}$-supporting dynamic changes of workflows without losing control. Int. J. Intell. Inf. Syst. (JIIS) **10**(2), 93–129 (1998)
395. Reichert, M., Rinderle, S., Dadam, P.: ADEPT workflow management system. In: Business Process Management, pp. 370–379. Springer, Berlin (2003)
396. Reijers, H.A., Rigter, J.H.M., van der Aalst, W.M.P.: The case handling case. Int. J. Cooperative Inf. Syst. **12**(3), 365–391 (2003)
397. Reimer, U., Margelisch, A., Novotny, B., Vetterli, T.: Eule2: A knowledge-based system for supporting office work. ACM SIGGROUP Bull. **19**(1), 56–61 (1998)
398. Reimer, U., Margelisch, A., Staudt, M.: Eule: A knowledge-based system to support business processes. Knowl.-Based Syst. **13**(5), 261–269 (2000)
399. Resende, L.: Handling heterogeneous data sources in a SOA environment with service data objects (SDO). In: Special Interest Group on Management of Data Conference, Beijing, pp. 895–897 (2007)
400. Riss, U., Rickayzen, A., Maus, H., van der Aalst, W.M.P.: Challenges for business process and task management. J. Universal Knowl. Manag. **0**(2), 77–100 (2005)
401. Romero, O., Abelló, A.: A survey of multidimensional modeling methodologies. Int. J. Data Warehouse. Min. **5**(2), 1–23 (2009)
402. Rozinat, A., van der Aalst, W.M.P.: Conformance checking of processes based on monitoring real behavior. Inf. Syst. **33**(1), 64–95 (2008)
403. Rozsnyai, S., Slominski, A., Lakshmanan, G.T.: Automated correlation discovery for semi-structured business processes. In: International Conference on Data Engineering Workshops, Hannover, pp. 261–266 (2011)
404. Ryman, A., Chinnici, R., Moreau, J.-J., Weerawarana, S.: Web Service Description Language (WSDL) Version 2.0. W3C Working Draft. http://www.w3.org/TR/wsdl20 (2007)
405. Ryndina, K., Küster, J.M., Gall, H.: Consistency of business process models and object life cycles. In: Models in Software Engineering, pp. 80–90. Springer, Heidelberg (2007)
406. Ryu, S.H., Benatallah, B.: Integrating feature analysis and background knowledge to recommend similarity functions. In: Proceedings of the 13th International Conference on Web Information Systems Engineering (WISE 2012), Paphos, 28–30 November 2012, pp. 673–680
407. Ryu, S., Benatallah, B., Paik, H., Kim, Y., Compton, P.: Similarity function recommender service using incremental user knowledge acquisition. In: International Conference on Service Oriented Computing (ICSOC), Paphos (2011)
408. Sadiq, S., Orlowska, M.: Analyzing process models using graph reduction techniques. Inf. Syst. **25**, 117–134 (2000)
409. Sakr, S., Al-Naymat, G.: Relational processing of RDF queries: A survey. SIGMOD Rec. **38**(4), 23–28 (2009)
410. Sakr, S., Awad, A.: A framework for querying graph-based business process models. In: Proceedings of the 19th International Conference on World Wide Web (WWW), Raleigh, pp. 1297–1300 (2010)
411. Salton, G., Wong, A., Yang, C.S.: A vector space model for automatic indexing. Commun. ACM **18**(11), 613–620 (1975)
412. SAP AG: SAP Business Intelligence. http://www.sap.com/platform/netweaver/pdf/BWP_BI_Overview.pdf (2005)
413. Sarma, A.D., Dong, X.L., Halevy, A.Y.: Data modeling in dataspace support platforms. In: Conceptual Modeling: Foundations and Applications, pp. 122–138. Springer, Heidelberg (2009)
414. Satish, A., Jain, R., Gupta, A.: Tolkien: An event based storytelling system. Proc. VLDB Endowment **2**, 1630–1633 (2009)
415. Scheer, A.-W., Hoffmann, M.: The process of business process management. Handb. Bus. Process Manag. (2), 351–380 (2015)
416. Schwalbe, K.: Information Technology Project Management, Revised. Cengage Learning, Boston (2010). http://muele.mak.ac.ug/pluginfile.php/200806/mod_resource/content/1/Book%20Bhwalbe.pdf

417. Sebahi, S., Hacid, M.-S.: Business process monitoring with BPath. In: On the Move to Meaningful Internet Systems (OTM), pp. 446–453. Springer, Heidelberg (2010)
418. Sellis, T., Lin, C.-C., Raschid, L.: Coupling production systems and database systems: A homogeneous approach. IEEE Trans. Knowl. Data Eng. **5**(2), 240–256 (1993)
419. Shapiro, L.G., Haralick, R.M.: Structural descriptions and inexact matching. IEEE Trans. Pattern Anal. Mach. Intell. **3**, 504–519 (1981)
420. Shvaiko, P., Euzenat, J.: A survey of schema-based matching approaches. J. Data Semantics **IV**, 146–171 (2005)
421. Simmhan, Y., Plale, B., Gannon, D.: A survey of data provenance in e-science. SIGMOD Rec. **34**(3), 31–36 (2005)
422. Skersys, T., Tutkute, L., Butleris, R., Butkiene, R.: Extending bpmn business process model with sbvr business vocabulary and rules. Inf. Technol. Control **41**(4), 356–367 (2012)
423. Srikanth, I., Behara, G.K.: Service identification: BPM and SOA handshake. BPTrends 3: White paper (2007). http://www.bitpipe.com/detail/RES/1193674938_452.html
424. Staab, S., Schnurr, H.-P.: Smart task support through proactive access to organizational memory. Knowl.-Based Syst. **13**(5), 251–260 (2000)
425. Stoitsev, T., Scheidl, S., Spahn, M.: A framework for light-weight composition and management of ad-hoc business processes. In: Proceedings of 6th International Workshop on Task Models and Diagrams for User Interface Design (TAMODIA 2007), Toulouse, 7–9 November 2007. Lecture Notes in Computer Science, vol. 4849, pp. 213–226. Springer, Berlin (2007) [ISBN 978-3-540-77221-7]
426. Stonebraker, M.: SQL databases v. NoSQL databases. Commun. ACM **53**(4), 10–11 (2010)
427. Strosnider, J.K., Nandi, P., Kumaran, S., Ghosh, S., Arsnajani, A.: Model-driven synthesis of soa solutions. IBM Syst. J. **47**(3), 415–432 (2008)
428. Su, S.Y.W., Huang, C., Hammer, J., Huang, Y., Li, H., Wang, L., Liu, Y., Pluempitiwiriyawej, C., Lee, M., Lam, H.: An internet-based negotiation server for e-commerce. VLDB J. **10**(1), 72–90 (2001)
429. Subramanian, S., Thiran, P., Narendra, N.C., Mostéfaoui, G.K., Maamar, Z.: On the enhancement of BPEL engines for self-healing composite Web services. In: Proceedings of the 2008 International Symposium on Applications and the Internet (SAINT 2008), Turku, 28 July–1 August 2008, pp. 33–39. IEEE Computer Society, Los Alamitos (2008) [ISBN 978-0-7695-3297-4]
430. Sun, Y., Han, J., Zhao, P., Yin, Z., Cheng, H., Wu, T.: RankClus: Integrating clustering with ranking for heterogeneous information network analysis. In: Proceedings of 12th International Conference on Extending Database Technology (EDBT 2009), Saint Petersburg, 24–26 March 2009. ACM International Conference Proceeding Series, vol. 360. ACM, New York (2009) [ISBN 978-1-60558-422-5]
431. Suwannopas, P., Senivongse, T.: Discovering semantic web services with process specifications. In: Proceedings of 6th IFIP WG 6.1 International Conference on Distributed Applications and Interoperable Systems (DAIS 2006), Bologna, 14–16 June 2006. Lecture Notes in Computer Science, vol. 4025, pp. 113–127. Springer, Berlin (2006) [ISBN 3-540-35126-4]
432. Swenson, K., Fischer, L., Kemsley, S., Palmer, N.L., Richardson, C.: Social BPM: Work, Planning and Collaboration Under the Impact of Social Technology. Bpm and Workflow Handbook Series. CreateSpace Independent Publishing Platform, USA (2011) [CreateSpace is a DBA of On-Demand Publishing LLC, part of the Amazon group of companies]
433. Swenson, K.D., Palmer, N., Silver, B.: Taming the Unpredictable Real World Adaptive Case Management: Case Studies and Practical Guidance. Future Strategies Inc., Lighthouse Point (2011)
434. Sycara, K.P., Paolucci, M., Ankolekar, A., Srinivasan, N.: Automated discovery, interaction and composition of semantic web services. J. Web Semantic **1**(1), 27–46 (2003)
435. Sycara, K., Klusch, M., Fries, B.: Automated sematic web discovery with owls-mx. In: Proceedings of the Fifth International Joint Conference on Autonomous Agents and Multiagent Systems, Hakodate (2006)

436. Syeda-Mahmood, T.F., Shah, G., Akkiraju, R., Ivan, A.-A., Goodwin,R.: Searching service repositories by combining semantic and ontological matching. In: ICWS, IEEE International Conference on Web Services, 11–15 July 2005, Orlando, pp.13–20 (2005)

437. Tailor: A record linkage tool box. In: Proceedings of the 18th International Conference on Data Engineering, ICDE '02. IEEE Computer Society, Los Alamitos (2002)

438. Tee, K., Greenberg, S., Gutwin, C.: Providing artifact awareness to a distributed group through screen sharing. In: Proceedings of the 2006–20th Anniversary Conference on Computer Supported Cooperative Work, pp. 99–108. ACM, New York (2006)

439. ter Hofstede, A.H.M., Ouyang, C., La Rosa, M., Song, L., Wang, J., Polyvyanyy, A.: APQL: A process-model query language. In: Selected Papers. Asia Pacific Business Process Management - First Asia Pacific Conference (AP-BPM 2013), Beijing, 29–30 August 2013. Lecture Notes in Business Information Processing, vol. 159. Springer, Berlin (2013) [ISBN 978-3-319-02921-4]

440. Thiagarajan, R.K., Srivastava, A.K., Pujari, A.K., Bulusu, V.K.: BPML: A process modeling language for dynamic business models. In: Fourth IEEE International Workshop on Advanced Issues of E-Commerce and Web-Based Information Systems (WECWIS'02/WECWIS 2002), Newport Beach, 26–28 June 2002, pp. 239–241

441. Thomsen, E.: OLAP Solutions: Building Multidimensional Information Systems, 2nd edn. Wiley, New York (2002)

442. Thullner, R., Rozsnyai, S., Schiefer, J., Obweger, H., Suntinger, M.: Proactive business process compliance monitoring with event-based systems. In: Workshops Proceedings of the 15th IEEE International Enterprise Distributed Object Computing Conference (EDOCW 2011), Helsinki, 29 August–2 September 2011, pp. 429–437. IEEE Computer Society 2011 [ISBN 978-1-4577-0869-5]

443. Tosic, V., Pagurek, B., Patel, K.: Wsol - a language for the formal specification of classes of service for web services. In: International Conference on Web Services (ICWS), Las Vegas, pp. 375–381 (2003)

444. Trkman, P., McCormack, K., de Oliveira, M.P.V., Ladeira, M.B.: The impact of business analytics on supply chain performance. Decis. Support Syst. **49**(3), 318–327 (2010)

445. Truong, H.L., Dustdar, S.: On analyzing and specifying concerns for data as a service. In: Asia-Pacific Services Computing Conference (APSCC), Singapore, pp. 87–94 (2009)

446. Türetken, O., Elgammal, A., van den Heuvel, W.-J., Papazoglou, M.P.: Enforcing compliance on business processes through the use of patterns. In: 19th European Conference on Information Systems (ECIS), Helsinki, 9–11 June 2011 (2011)

447. Vaculín, R., Sycara, K.P.: Towards automatic mediation of owl-s process models. In: 2007 IEEE International Conference on Web Services (ICWS), 9–13 July 2007, Salt Lake City, pp. 1032–1039 (2007)

448. van der Aalst, W.M.P., Pesic, M.: DecSerFlow: Towards a Truly Declarative Service Flow Language. Springer, Heidelberg (2006)

449. van der Aalst, W.M.P., ter Hofstede, A.H.M.: Yawl: Yet another workflow language. Inf. Syst. **30**(4), 245–275 (2005)

450. van der Aalst, W.M.P., van Hee, K.M.: Workflow Management: Models, Methods, and Systems. MIT, Cambridge (2002)

451. van Dongen, B.F., Dijkman, R., Mendling, J.: Measuring similarity between business process models. In: Proceedings of 20th International Conference on Advanced Information Systems Engineering (CAiSE 2008), Montpellier, 16–20 June 2008. Lecture Notes in Computer Science, vol. 5074. Springer, Berlin (2008) [ISBN 978-3-540-69533-2]

452. van Glabbeek, R.J.: The linear time - branching time spectrum ii. In: 4th International Conference on Concurrency Theory (CONCUR '93), Hildesheim, 23–26 August. Lecture Notes in Computer Science, vol. 715, pp. 278–297. Springer, Heidelberg (1990)

453. Van Hentenryck, P., Saraswat, V.A.: Strategic directions in constraint programming. ACM Comput. Surv. **28**(4), 701–726 (1996)

454. Vanderfeesten, I., Reijers, H.A., van der Aalst, W.M.P.: Product Based Workflow Design with Case Handling Systems. BETA Working Paper Series, WP 189. Eindhoven University of Technology, Eindhoven (2006)

455. Vanthienen, J., Goedertier, S.: How business rules define business processes. Bus. Rules J. **8**(3) (2007)

456. Vasilecas, O., Lebedys, E.: Application of business rules for data validation. Inf. Technol. Control **36**(3), 273–277 (2007)

457. von Ammon, R.: Event-driven business process management. In: Encyclopedia of Database Systems, pp. 1068–1071. Springer, Berlin (2009)

458. von Halle, B., Ronald, G.: Business Rules Applied: Building Better Systems Using the Business Rules Approach. Wiley, New York (2001)

459. Wang, L., Ranjan, R., Chen, J., Benatallah, B.: Cloud Computing: Methodology, Systems, and Applications. CRC Press, Taylor and Francis Group (2012). https://www.crcpress.com/Cloud-Computing-Methodology-Systems-and-Applications/Wang-Ranjan-Chen-Benatallah/9781439856413

460. Wang, J., Jin, T., Wong, R.K., Wen, L.: Querying business process model repositories - A survey of current approaches and issues. World Wide Web **17**(3), 427–454 (2014)

461. Wang, Z., Wen, L., Wang, J., Wang, S.: TAGER: transition-labeled graph edit distance similarity measure on process models. In: On the Move to Meaningful Internet Systems: OTM 2014 Conferences. Proceedings of the Confederated International Conferences: CoopIS, and ODBASE 2014, Amantea, 27–31 October 2014. Lecture Notes in Computer Science, vol. 8841, pp. 184–201 (2014)

462. Wargitsch, C., Wewers, T., Theisinger, F.: An organizational-memory-based approach for an evolutionary workflow management system-concepts and implementation. In: Proceedings of the Thirty-First Hawaii International Conference on System Sciences, vol. 1, pp. 174–183. IEEE, New York (1998)

463. Watts, D.J.: Networks: An Introduction. Princeton University Press, Princeton (2003)

464. Weber, B., Wild, W.: Towards the agile management of business processes. In: WM 2005: Contributions to the 3rd Conference Professional Knowledge Management - Experiences and Visions, 10–13 April 2005, Kaiserslautern, pp. 409–419. DFKI, Kaiserslautern (2005)

465. Weber, I., Paik, H.-Y., Benatallah, B.: Form-based web service composition for domain experts. ACM Trans. Web (TWEB) **8**(1), 2 (2013)

466. Weidlich, M., Dijkman, R.M., Mendling, J.: The icop framework: Identification of correspondences between process models. In: Proceedings of the 22nd International Conference on Advanced Information Systems Engineering (CAiSE), Hammamet, 7–9 June 2010, pp. 483–498 (2010)

467. Weidlich, M., Ziekow, H., Mendling, J., Günther, O., Weske, M., Desai, N.: Event-based monitoring of process execution violations. In: Business Process Management (BPM), pp. 182–198. Springer, Heidelberg (2011)

468. Weidlich, M., Sagi, T., Leopold, H., Gal, A., Mendling, J.: Predicting the quality of process model matching. In: 11th International Conference on Business Process Management (BPM), Beijing, pp. 203–210 (2013)

469. Weske, M.: Business Process Management: Concepts, Languages, Architectures. Springer, Berlin (2007)

470. White, T.: Hadoop: The Definitive Guide, 1st edn. O'Reilly Media, Sebastopol (2009)

471. Whittaker, S., Frohlich, D., Daly-Jones, O.: Informal workplace communication: What is it like and how might we support it? In: Proceedings of the SIGCHI Conference on Human Factors in Computing Systems, pp. 131–137. ACM, New York (1994)

472. Widom, J.: Trio: A system for integrated management of data, accuracy, and lineage. In: Conference on Innovative Data Systems Research (CIDR), Asilomar, pp. 262–276 (2005)

473. Wiederhold, G., Wegner, P., Ceri, S.: Toward megaprogramming. Commun. ACM **35**(11), 89–99 (1992)

474. Wombacher, A., Fankhauser, P., Mahleko, B., Neuhold, E.J.: Matchmaking for business processes based on choreographies. Int. J. Web Serv. Res. **1**(4), 14–32 (2004)

475. Wu, Z., Palme, M.S.: Verb semantics and lexical selection. In: ACL-32nd Annual Meeting of the Association for Computational Linguistics, pp. 133–138. Morgan Kaufmann, Los Altos (1994)

476. Yaeli, A., Kofman, A., Dubinsky, Y.: Software development governor: Automating governance in software development environments. In: 31st International Conference on Software Engineering: Companion Volume (ICSE-Companion), pp. 413–414. IEEE, New York (2009)

477. Yan, Z., Dijkman, R.M., Grefen, P.W.P.J.: Fnet: An index for advanced business process querying. In: Proceedings of the 10th International Conference on Business Process Management (BPM), Tallinn, 3–6 September 2012, pp. 246–261 (2012)

478. Yuan, Y., Lin, X., Liu, Q., Wang, W., Yu, J.X., Zhang, Q.: Efficient computation of the skyline cube. In: Proceeding of the Very Large Data Bases, Trondheim, pp. 241–252 (2005)

479. Zaremski, A.M., Wing, J.M.: Signature matching: A tool for using software libraries. ACM Trans. Softw. Eng. Methodol **4**(2), 146–170 (1995)

480. Zaremski, A.M., Wing, J.M.: Specification matching of software components. ACM Trans. Softw. Eng. Methodol. **6**(4), 333–369 (1997)

481. Zhao, K., Ying, S., Zhang, L., Hu, L.: Achieving business process and business rules integration using spl. In: International Conference on Future Information Technology and Management Engineering (FITME), vol. 2, pp. 329–332. IEEE, New York (2010)

482. Zhou, C., Chia, L.-T., Lee, B.-S.: Daml-qos ontology for web services. In: International Conference on Web Services (ICWS), San Diego, pp. 472–479 (2004)

483. Zikopoulos, P., Eaton, C.: Understanding Big Data: Analytics for Enterprise Class Hadoop and Streaming Data, 1st edn. McGraw-Hill Education, New York (2011)

Index

Activity-centric Process, 23
Apache Hadoop, 139
Apache HBase, 139
Apache Mahout, 142
Apache Pig, 139
Apache S4, 139
Apache Spark, 139
Apache ZooKeeper, 139
APQL, 98
Artifact centric, 37
Artifact-centric Process, 23

Beehivez, 97
Behavioral matching, 85
Big data analytics, 113
Big process graphs, 113
BP-Mon, 100
BP-QL, 96
BP-SPARQL, 101
BPEL, 96
BPM in the Cloud, 139
BPM lifecycle, 4
BPMN-Q, 93
Business process analysis (BPA), 5
Business process compliance, 102
Business process modeling notation (BPMN), 93
Business processes, 3
Business processes management, 3
Business rules approach (BRA), 41

Case handling, 8
Case management, 8, 56
Collaborative process, 57

Component model, 33
Constraint-based matching, 67
Content management system (CMS), 8
Correlation discovery, 10
Correlation syntax, 35
CRM, 8
Customer relationship management (CRM), 8

Data as a service (DaaS), 12
Data centric, 37
Data service, 109
Dataspace, 12, 110
DataSpace Support Platform (DSSPs), 12, 110
Delta analysis, 64
Deontic rules, 42
Derivation rules, 42

ECA rules, 26
Event driven business process management (EDBPM), 45
Event-Condition-Action Rules, 26
Exception handling syntax, 34
Execution engine, 24
Extract, transform, and load (ETL), 108

Finite state machine (FSM), 78

Hadoop, 13, 139
HDFS, 139

Integrity rules, 42
Interface matching, 69

© Springer International Publishing Switzerland 2016
S.-M.-R. Beheshti et al., *Process Analytics*, DOI 10.1007/978-3-319-25037-3